PROGRESS IN FIELDWORK

GEOGRAPHY

HAYLEY PEACOCK

KEY STAGE 3

T0187393

HODDER
EDUCATION
AN HACHETTE UK COMPANY

The Publishers would like to thank the following for permission to reproduce copyright material.

Photo credits

p.2 © magicbones - stock.adobe.com; **p.3** © Greg Balfour Evans / Alamy Stock Photo; **p.5** © Greg Balfour Evans / Alamy Stock Photo; **p.7** © In Pictures Ltd./Corbis via Getty Images; **p.12** r https://www.oponeo.co.uk/blog/on-the-road-to-greener-future; **p.13** © Eliza - stock.adobe.com ; **p.17** © zhang yongxin - stock.adobe.com; **p.24** t © Maddie Red Photography / Alamy Stock Photo, b © mypuy - stock.adobe.com; **p.26** tl © Rawpixel.com - stock.adobe.com, bl © Trevor Mogg / Alamy Stock Photo, tr © Matthew Chattle / Alamy Stock Photo, br © LH Images / Alamy Stock Photo; **p.29** © F. Krawen - stock.adobe.com; **p.31** t © Robert Evans / Alamy Stock Photo, bl © Greg Balfour Evans / Alamy Stock Photo, br © Parmorama / Alamy Stock Photo; **p.37** © Ben Schonewille -stock.adobe.com; **p.38** © A.P.S. (UK) / Alamy Stock Photo; **p.39** bl © Aleksandr Ugorenkov - stock.adobe.com; **p.40** t © Aleksandr Ugorenkov - stock.adobe.com, bl © ansala - stock.adobe.com, br © Patrick Daxenbichler - stock.adobe.com; **p.41** © Stéphane Lecointre - stock.adobe.com; **p.46** t © Nik - stock.adobe.com, b © wildman.photos - stock.adobe.com; **p.47** © George Little/Associated Newspapers/Shutterstock; **p.52** © cheekylorns - stock.adobe.com; **p.53** © phil_bird/123RF; **p.56** © Rusana - stock.adobe.com; **p.57** © John Ewing/Portland Press Herald via Getty Images; **p.59** © lcswart - stock.adobe.com; **p.60** © andrewmroland - stock.adobe.com; **p.65** tr © Steven May / Alamy Stock Photo, br © Steven May / Alamy Stock Photo, tl © a-plus image bank / Alamy Stock Photo, tc © lemanieh - stock.adobe.com; **p.66** © Steven May / Alamy Stock Photo; **p.67** © a-plus image bank / Alamy Stock Photo; **p.68** © tonefotografia - stock.adobe.com; **p.70** © Hasan - stock.adobe.com; **p.72** © Sevenstock Studio - stock.adobe.com; **p.85** © Nadia Snopek / Shutterstock.com; **p.95** © Spencer Platt/Getty Images; **p.97** © Jobakashii - stock.adobe.com; **p.98** © kaliantye - stock.adobe.com; **p.103** © pavel siamionov - stock.adobe.com; **p.104** © Gary Perkin - stock.adobe.com

Acknowledgements and text credits

With special thanks to John Widdowson.

We are grateful to the geography department at Woodford County High School for the fieldwork ideas that inspired Chapter 5.

p.61 and **p.65** contains Environment Agency data © Crown Copyright and © Crown Copyright Ordnance Survey; **p.62** River Don at Doncaster, © River Levels UK. This page also contains Environment Agency data © Crown Copyright; **p.65** River Wye at Hereford Bridge, © River Levels UK. This page also contains Environment Agency data © Crown Copyright

Orders: please contact Hachette UK Distribution, Hely Hutchinson Centre, Milton Road, Didcot, Oxfordshire, OX11 7HH. Telephone: +44 (0)1235 827827. Email education@hachette.co.uk. Lines are open from 9 a.m. to 5 p.m., Monday to Friday. You can also order through our website: www.hoddereducation.com

ISBN: 978 1 5104 7756 8

© Hayley Peacock 2020

First published in 2020 by
Hodder Education,
An Hachette UK Company
Carmelite House
50 Victoria Embankment
London EC4Y 0DZ
www.hoddereducation.co.uk

Impression number   10 9 8 7 6 5 4 3 2

Year   2024 2023

Cover photo © arquiplay77 – stock.adobe.com

Illustrations by DC Graphics Design Limited

Typeset in Museo Sans by DC Graphic Design Limited, Hextable, Kent

Printed and bound by CPI Group (UK) Ltd, Croydon, CR0 4YY

A catalogue record for this title is available from the British Library.

Contents

Contents

Introduction

Why do we do fieldwork?

Put simply, fieldwork is *doing* geography. Rather than reading all about places and landscapes in a book, fieldwork allows you to study geography outside in the real world. Fieldwork gives you the chance to actively explore the world around you, research it, and collect information or data on a specific topic to understand it. It might take you to expansive coastlines, quiet neighbourhoods or bustling urban areas. You might need to use special instruments to collect your data, or just a pen and paper.

This fieldwork book gives you opportunities to study your local area. In order to do this, you will use a range of fieldwork techniques to investigate the world around you. You might find that you aren't able to go out and collect the data you need – don't panic! In this book, there are plenty of example datasets to use, and chances to explore places virtually.

Who uses fieldwork?

A conservationist uses fieldwork to help them replenish trees in danger of being lost.

A meteorologist uses fieldwork to assess the weather and alert people if there is severe weather on the way.

A geologist uses fieldwork to work out ways to prevent floods or earthquakes.

An urban geographer uses fieldwork to examine how cities are changing.

An environmentalist uses fieldwork to assess the impact climate change is having.

A social researcher uses fieldwork to gather evidence on a social issue, such as homelessness.

You can even do fieldwork on the moon! Astronauts have received fieldwork training so they can examine what the moon is made of.

How to use this book

All fieldwork enquiries in this book follow a structure:

Preparation
↓
Collection
↓
Presentation
↓
Analysis
↓
Evaluation

Preparation: This stage is how all fieldwork begins. It starts with a key enquiry question, which is the question that you will be researching and seeking to answer. Good preparation will involve getting some context to your enquiry, to help you understand the topic better.

Collection: In this stage, you will design your fieldwork techniques and use them to collect your data. These will be the sheets you take out to your fieldwork sites. Examples are surveys, assessments and questionnaires.

Presentation: After you've collected your data, you need to make sense of it. This stage will give you different ways of presenting your data in charts, graphs, maps and posters.

Analysis: Using your presented data, this next stage will take you through how to interpret what you have found. This might mean looking at trends in graphs, patterns in questionnaire answers or spotting unusual results.

Evaluation: Evaluation means reflecting on the problems you experienced during your fieldwork. In this stage, you will be taken through some questions that will help you think about what went well in your fieldwork, and what didn't.

It's time to begin your first fieldwork enquiry!

Learning objectives

▶ To prepare for a geographical enquiry on cycle safety.

▶ To create a traffic count.

▶ To design a cycling questionnaire.

▶ To present findings in a bar chart and a pie chart.

▶ To analyse and evaluate my enquiry.

Why is cycle safety important?

Cycling is a popular, healthy way to get from A to B. Cycling also helps reduce traffic congestion because more people cycling to work means fewer people driving to work. In countries like the Netherlands, famous for its cycle-friendly streets, almost a third of all journeys are by bike. Only two per cent of the UK population, by contrast, commute by bike. Perhaps this is partly because UK cycling can be dangerous: in London alone, around 14 cyclists die in road accidents each year. These dangers are also known as hazards – something that could cause harm. For example, a pothole in the road is a hazard, because if a cyclist rides over one, they could lose their balance and fall. In order to get more people to commute by bike, it is important to first make sure cycling is safe. In a pledge to try and get more people on their bikes, the UK Government has promised to increase cycle safety to encourage more people to cycle rather than use their cars.

Activity

1 List five benefits of cycling and five dangers of cycling. Use the text above to help you, as well as your own knowledge and experience.

Benefits	Dangers
1 It reduces traffic congestion.	1 Potholes can make you fall off your bike.
2	2
3	3
4	4
5	5

Designing the question for your enquiry

All geographical enquiries need a key enquiry question. You must consider this question throughout the fieldwork and use the information, or data, you have gathered over the course of your investigation to answer it. The data you will collect is called primary data, as it is new information collected by yourself during fieldwork.

Thinking about cycle safety, our enquiry question here will be: How safe do people think my local area is for cyclists?

Ⓐ Only two per cent of the UK population commute by bike

Activities

2 Before you begin your own fieldwork enquiry, imagine you are a cyclist travelling through a city of your choice. Put on your imaginary cycle helmet and find your city in Google Maps. Select Street View and use the white arrows to explore the area by going up and down the roads:

a) Identify the hazards you discover during your virtual trip.

b) Give examples of cycle safety that you see during your virtual cycle trip.

c) Now type in your school name in the Google Maps search bar. Think about your journey to school and find a busy road that might be used by cyclists. This would be a good place to begin your enquiry.

d) Where is the road you are thinking of located? Is it north, south, east or west of school?

3 a) Look at Figure B and the example annotations provided. The annotations are in pairs. For each pair, tick the one annotation that you think describes the image correctly.

b) Write down three ways you might make this road safer for cyclists.

☐ Plenty of space for a car door to open without hitting a cyclist

☐ Side door dangerous for passing cyclists

☐ Smooth surface to cycle on

☐ Cycle path damaged

☐ Clearly outlined cycle lane

☐ Cycle lane shared with pedestrians

☐ Separate cycle lane away from road

☐ Cyclists must share with other road users

☐ Plenty of cycle racks provided

☐ Not enough cycle racks to meet demand

B A main road in Haringey, London

Collecting data

For this enquiry, a traffic count and a questionnaire will help you look at the factors that make cycling safe or unsafe in your local area. If you are going to conduct your fieldwork as a class, your teacher will discuss with you the best road or junction for you to study. Alternatively, you could complete the activities using data recorded on the A1018 road in Sunderland – a city in the north-east of England (Figure C).

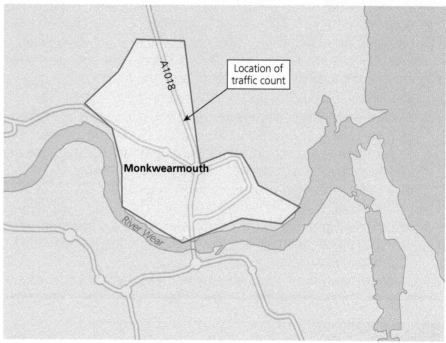

C The location of the Sunderland traffic count

Activities

4 a) What kinds of vehicles might you expect to see during your traffic count? Use the template in Figure D to record your answers.

   b) Look at Photo E on the next page: the photograph of the students collecting their traffic count data. List two ways that the students are being safe during their fieldwork. Can you find one way they could be even safer?

5 Using the step-by-step information on the next page, create a list of all the equipment you would need to conduct a traffic count.

Fieldwork technique 1: traffic count

Traffic counts record the number and type of vehicle using a road over a five-minute period.

During your virtual cycling trip, you may have noticed different amounts of traffic. A quiet, peaceful road may be safer for cyclists, while a busy road with lots of large vehicles can be quite frightening to cycle through. To investigate how busy roads are, geographers use traffic counts. Traffic counts can give you an idea of how safe that road might be for cyclists. For instance, a road that has plenty of big vehicles going through it might be less safe for cyclists than one with smaller vehicles like motorbikes.

Location:		Road:		
Date:		Time:		
Type of vehicle	Travelling left–right	Travelling right–left	Total number	
Cars				

D Traffic count template

E Two women collecting traffic count data

Step 1

If you are collecting your traffic count data in your local area, it is important to find a good, safe place away from the kerb and out of the way of pedestrians. You may wish to do different sections along the same road. Before you start your timer and get going, write the name of the road and its location on the top of the sheet, as well as the time of day.

Split yourselves into groups of two. Staying on the same side of the road, one of you will count and classify traffic going from left to right, and the other will count and classify traffic going right to left.

Step 2

Using a countdown timer, set yourselves five minutes to begin counting and classifying the traffic on the road. You will be looking for the number of cars, lorries, vans, buses, taxis, motorbikes and bicycles. Every time you see a vehicle that matches the one on your sheet, mark a small line in the box and tally up your score. This makes it easier to count up when your five minutes is over!

Step 3

Once your time is up, everyone must stop at the same time. Using only your own sheet, count up the total number of each type of vehicle you counted on your side of the road. Check with your partner what scores they got for the opposite direction of traffic, and fill in their results onto the last column on your sheet. You should now have data for traffic going left to right, and from right to left. At the end, work out the total number of vehicles you counted, and add the total score to the final column on your sheet.

Location: Sunderland		Road: A 1808	
Date: 17th May, 2020		Time: 2:15pm	
Type of vehicle	Travelling left–right	Travelling right–left	Total number
Cars	III	IIII	IIII III

F Traffic count template with some example data

Fieldwork technique 2: questionnaire

A questionnaire is a way of gathering people's responses to the same set of questions. A questionnaire is a good way of gathering interesting primary data, and is a very useful fieldwork technique in geography. Other people's opinions provide valuable insights into a topic and can be very revealing.

To investigate how safe your area is for cyclists, you will create a cycling questionnaire to ask specific questions about the reasons why people do or do not cycle, and what would make them change their habits.

Activity

6 Before going out into the field, you need to design your questionnaire. It's important that the questionnaire is not too long so that you don't take up too much of your respondents' time. It's also quite difficult to do correctly, so it needs careful thinking about.

a) Below is a list of possible questions you could include in your questionnaire. Mix and match the following questions on the left with these possible answer choices on the right:

Do you own your own bike?	Less traffic/More cycle lanes/Cheaper bike hire/Owning own bike
How often do you use a bike?	Car/Bus/Train/Bike/Walk
What transport do you use for short journeys (less than a mile)?	Every day/Every week/Less than once a week/Never
What transport do you use for long journeys (over a mile)?	Car/Bus/Train/Bike/Walk
What would encourage you to ride a bike more often?	Yes/No and don't want to/No but would like to
Do you feel safe cycling here?	Yes/No/Sometimes

G Possible questionnaire questions and answer choices

b) On a new page or piece of paper, write a section at the top of your questionnaire for 'location', 'date' and 'time'. Why are these pieces of information important to record on fieldwork sheets?

c) Design your questionnaire so that you can easily record people's responses. Using the mix and match answers from activity 6a) plus the example below to help you, add in the questions you want people to answer and the answer options that will make it quick and easy to record their responses. Use this example to help you come up with questions and answer options. There are blanks left in it so you think about the right questions and answer options for your area!

Questions	Answer options			
1 Do you own your own bike?	Yes	No and don't want to	No but would like to	
2 How often do you use a bike?	Every day			
3	Less traffic	More cycle lanes	Cheaper bike hire	Owning own bike
4 Do you feel safe cycling here?				
5				
6				

H Questionnaire template

❶ Students asking questions from a questionnaire

When you have designed your questionnaire, it's time to go out into the field.

Step 1

Get yourselves into small groups of two or three. Everyone should have a copy of a questionnaire at hand. It's really important to stay in your groups, and to listen to your teacher's instructions about how far to go. Safety is key!

Step 2

Now is the fun part! You need to approach someone and see if they will stop for a couple of minutes to speak to you. Try to be as specific as possible when you approach someone. Avoid saying, 'Excuse me, can I have a couple of minutes of your time?': people might wonder if you are trying to sell something and be put off! Instead, try this:

> Excuse me, we are conducting a geography enquiry for school on cycling and wondered if you wouldn't mind answering some short questions?

Step 3

Start with your first question. Mark the response in the correct box. Ensure everyone in your group marks their responses. When you have finished your questions, make sure you always say thank you! Even if you get turned away by someone, remember that it's not personal and they are probably just very busy. Always be polite.

Data presentation is a useful way of interpreting the information you have collected through your traffic count and questionnaire by making it easier to read. There are many ways of presenting your data in geography. A good place to start is to consider whether the data you have is quantitative or qualitative. Quantitative data involves numbers or statistics and is often presented in charts. Qualitative data collects information that is not numerical. Instead, the data collected is descriptive. Interviews are a good example of qualitative data, as people can describe answers in their own words.

Bar chart

A bar chart is one data presentation technique in geographical fieldwork. It is used to show the frequency (count) data in vertical blocks or bars.

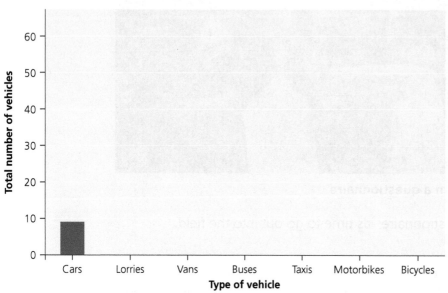

J **Traffic count in Sunderland**

Use this example of a bar chart to help you when creating your own.

Activities

7 What kind of data is used in the traffic count?

8 a) Count the total number of vehicles observed during the traffic count. To do this, add together the number of a type of vehicle going left to right, and going right to left.

b) On graph paper, create your x-axis. This is the line that runs horizontally across the bottom part of the graph. Write the types of vehicle from your traffic count along the bottom of the line, giving each vehicle a 1 cm box. Leave another 1 cm space between each vehicle on the x-axis, or your bars will be touching one another! Label this axis as 'Type of vehicle'.

c) Create your y-axis, which is the line that runs up the left-hand side vertically. This line will be used to show the number of each vehicle counted. Use the highest number in the 'total number' column to help you with how high to go on your y-axis, and how much you want to increase each value by, e.g. 2, 4, 6, 8... or 5, 10, 15, 20... or 10, 20, 30, 40... Label this axis 'Total number of vehicles'.

d) Using the data provided for Sunderland (Figure L), or the data you collected in your own traffic count, create the bars for your chart. Use the y-axis to help you work out how high to draw each bar for each vehicle.

Pie chart

A pie chart is used to show how data is split into proportions. In this geographical enquiry, we can present the data from the questionnaire in a pie chart to show the number of people that responded in particular ways on our questionnaire. For this, you will need a calculator, a compass and a protractor.

Do you own your own bike?

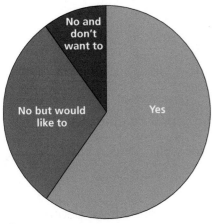

K Pie chart

Activity

9 To create a section in a pie chart, you need to work out some proportions and then angles. This will allow you to calculate the size of each segment in the pie chart.

a) Using the first question in the questionnaire, 'Do you own your own bike?', count the number of responses received for 'Yes', for 'No and don't want to' and for 'No but would like to'. For example, in the example given, the total would be 12 + 2 + 6.

b) Now calculate the fraction for 'Yes' by dividing the number of responses for 'Yes' by the total responses to the questionnaire.

	Yes	No and don't want to	No but would like to
Do you own your own bike?	12	2	6

c) To make the fraction into a part of the pie, the next step is to multiply it by 360. This number (360) is the number of degrees in a full circle, which is the shape of a pie chart.

$12/20 = 0.6$

$0.6 \times 360 = \boxed{}$

d) Now you have calculated the size of the slice for the 'Yes' answer, follow the same process for the two 'No...' responses. The number you get will be a number of degrees, which you use to make each slice in the pie chart.

$12/20 = \boxed{}$ $2/20 = \boxed{}$ $6/20 = \boxed{}$

e) On your paper, draw a big circle with your compass. Using your protractor, mark the degrees you calculated for 'Yes', 'No and don't want to' and 'No but would like to'. With each point, follow the line up into the middle of the pie. Colour in each slice and label them so it's clear which slice is which.

f) Continue this process for the other questions and results and create pie charts for each. Here are some data from the Sunderland traffic count that you could use:

Type of vehicle	Travelling left–right	Travelling right–left	Total number
Cars	13	44	13+44 = 57
Lorries	3	9	
Vans	16	8	
Buses	1	4	
Taxis	5	4	
Motorcycles	3	1	
Bicycles	2	0	

L Sunderland traffic count data

Good analysis is all about making sense of the geography that you have uncovered. It is a key part of all geographical enquiries and helps you to answer the key question set at the start. After analysing your data, the final step of any geographical enquiry is evaluating it. Evaluation is what geographers do after their analysis to look for some of the problems or issues they faced during their enquiry. Geography at GCSE and A Level is full of careful evaluation, so it's a great chance to start practising these skills now at KS3.

Analysing your data

Presenting your data in bar charts and pie charts makes it easier to analyse your results.

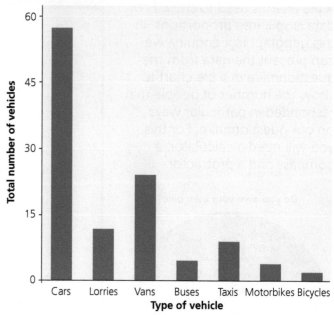

(M) **Results of traffic count, Sunderland**

Activities

10 Analyse the data in Figure M.
  a) Which vehicle was the most frequently recorded?
  b) Which vehicle was the least frequently recorded?
  c) What is the difference between the most and least frequently recorded vehicle?
  d) Think about the location used for the traffic count. Can you explain why there was such a high number of one type of vehicle on the road?

11 Using your pie chart data, copy out and complete the following paragraph of analysis, choosing the correct word or phrase from the options in parentheses, or by filling in the blank with the correct answer:

> The pie charts showed that most people *[do/do not]* cycle in this local area. In my investigation, I found that *[insert number of people]* people cycle, which was *[insert percentage]* of the total people asked in the questionnaire. Many people *[use their bike/take the bus/take the train/walk]* for short journeys.

> When asked what would encourage people to cycle more often, most people suggested that *[less traffic/more cycle lanes/cheaper bike hire/owning their own bike]* would help. Using the data collected in our fieldwork investigation, in conclusion I would argue that people in my local area *[do/do not/only sometimes]* feel safe cycling.

12 Using the passage above and your bar chart, complete the following sentences, choosing the appropriate word or phrase from the options provided in parentheses or filling in the blank:

> Our traffic count data *[supports/does not support]* our questionnaire data. During a traffic count, I found that there were *[insert number of most frequent vehicle]* compared to *[number of bicycles]*. To encourage more people to cycle safely, my local area should...

Evaluating your enquiry

Fieldwork is never perfect. There are often difficulties collecting data and also in making sense of what it means. As long as you can identify these problems (or limitations) then you can try and work out how they might affect your findings. Geographers call this evaluation, and it is an important part of any fieldwork enquiry.

Activity

13 a) Identify two limitations of your traffic counts that are linked to your equipment, including operator error (how it was used wrongly). Try and think about one limitation associated with the choice of equipment, and another about not using it correctly.

b) State a limitation of counting taxis. For example, think about whether they were always easy to identify.

c) Name two limitations of your questionnaire. For example, did you manage to find enough people to ask?

d) How did the time of day or day of the week affect your questionnaire? Where might most people be at that time of the day?

e) Identify the problem with the bar chart on the right (Figure N). How might you improve it?

f) Here are three pie charts: a, b and c (Figure O). Explain which of the three pie charts is the most appropriate, and why the other pie charts are not.

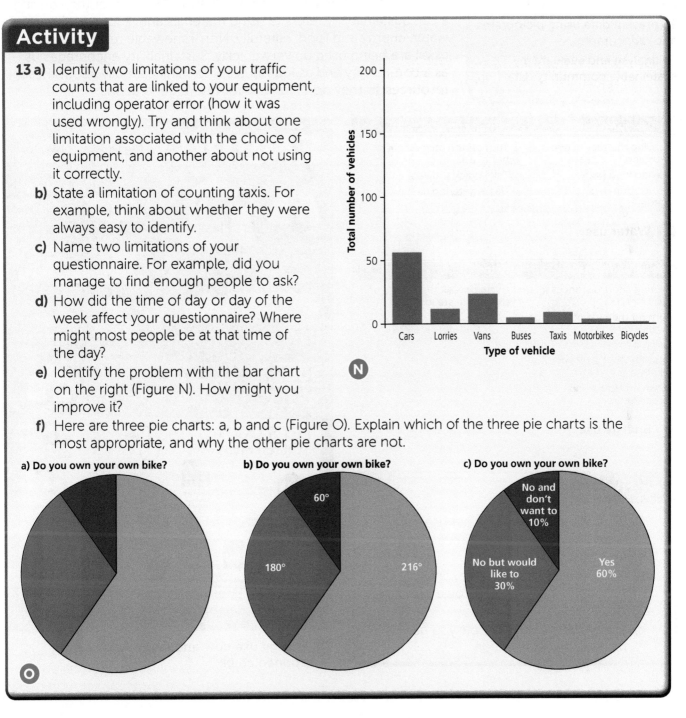

Congratulations! You have prepared, collected, presented, analysed and evaluated a geographical fieldwork enquiry.

Learning objectives

▶ To design a local sustainability survey.
▶ To create a sustainable communities assessment.
▶ To present data using pictograms and radar graphs.
▶ To analyse and evaluate my sustainable community data.

What is a sustainable community?

Sustainability is an important concept in geography. You may have heard the phrase 'being more sustainable' come up when people consider ways of looking after the planet. This may mean making some lifestyle changes, like riding a bike to work rather than driving a car. Sustainability refers to using materials and resources, like water, energy and food, carefully. Non-renewable resources such as oil are being used up very quickly. Sustainability encourages us – as a community and as individuals – to think carefully about these resources so they don't run out for future generations.

Not very sustainable	Better
Leaving the tap on when brushing your teeth	Turn off the tap when it is not in use
Having long baths	Take quick showers
Using sprinklers for the garden	Use a watering can

A Water use

Not very sustainable	Better
Driving short distances in a car	Cycle or walk the short distance instead
Turning the heating up high in the winter	
Using petrol in cars as fuel	
Leaving the TV on standby all night	
Washing clothes on a high temperature	

B Energy use

Activity

1 **a)** Read the examples of water use in Figure A. Complete the 'Better' column for energy use in Figure B.
   b) Using the examples, make some suggestions of how you could live more sustainably.
   c) Explain why each example you gave is a more sustainable option.

C Energy use now and in the future, from oponeo.co.uk

What makes a sustainable community?

Being sustainable is important for us as individuals, but it is most effective when whole communities live sustainably. How do you know if a community is a sustainable community or not? One way is to use the Egan Wheel, which is a local sustainability toolkit:

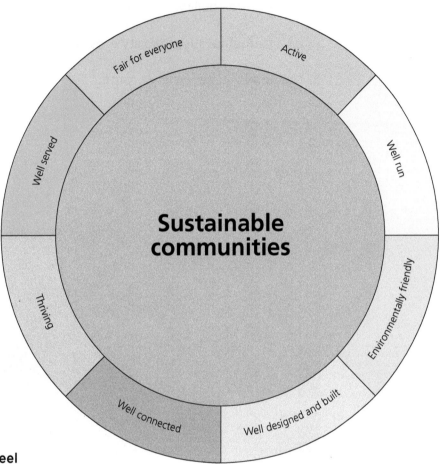

D The Egan Wheel

- Active – allowing people to exercise, cycle instead of drive, or walk safely.
- Well run – everyone gets to have a say on local decisions, e.g. whether a new car park is to be built on green space.
- Environmentally friendly – being kind to the planet.
- Well designed and built – such as new housing.
- Well connected – plenty of good transport links.
- Thriving – a successful local economy.
- Well served – plenty of schools and hospitals nearby.
- Fair for everyone – for instance, there are ramps for people in wheelchairs to get around, meaning they can access places that others may take for granted.

E Solar panels are an example of sustainability in a community

In geography, ideas are often grouped into three themes: social, environmental and economic. Sustainability can also be put into these three categories:

- social sustainability – people

- environmental sustainability – planet

- economic sustainability – profit.

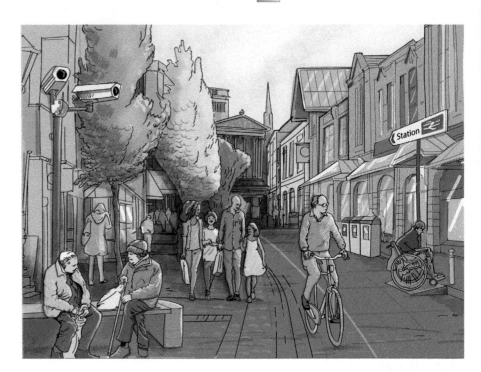

F **A sustainable community**

Activity

2 a) Classify the following ideas into social, economic and environmental. Give a reason for each of your choices.

Theme	Social, environmental or economic?
Active	Social, because it concerns people.
Well run	
Environmentally friendly	
Well designed and built	
Well connected	
Thriving	
Well served	
Fair for everyone	

b) Look at Figure F above. Find and list five examples of sustainability.

Although you have already thought about sustainability, doing a geographical enquiry on sustainability requires a more detailed approach. In the enquiry, you will use two fieldwork techniques to measure how sustainable a community is: a local sustainability survey and a sustainable community assessment. The local sustainability survey helps you find evidence of environmentally friendly activities or behaviours in your local area. The sustainable community assessment focuses on some of the key parts of the Egan Wheel.

Fieldwork technique 1: local sustainability survey

A local sustainability survey is like a treasure hunt, where you look for evidence of sustainability. An example of environmental sustainability might be if you see examples or signs of renewable energy being used, like solar panels on roofs or wind turbines in parks. You may also see public cycle hire, water bottle refilling stations or children cycling with their parents to school.

Activity

3 a) Create your local sustainability survey. You will need three columns: one for the evidence of sustainability you might see, one to let you tick or tally every time you see an example of it, and one to write down where you saw it.

b) You can use the examples in this template in your survey, but add three more examples of local sustainability you might see during your fieldwork.

Evidence	Tally	Where did I see it?
Cycle racks	II	Outside main school gate
Recycling bins		
Lights that turn off automatically		
Allotments for growing fruit, veg or herbs		
Green roofs		

G Local sustainability survey template

Step 1

In your local area (or in and around your school), spend 20 minutes exploring and looking for evidence of environmental sustainability with your teacher.

Step 2

Each time you find an example, mark it in the 'Tally' column.

Step 3

List each place you found it in under the third column; for example, 'In the dining room by the vending machine'.

Fieldwork technique 2: sustainable community assessment

You can use the Egan Wheel (Figure D on page 13) to create your own sustainable community assessment. By looking at each theme on the wheel, you can assess how sustainable that aspect of your local area (or school) is. Using what you see, smell and hear around you, you can mark how sustainable or not something is by using a scale from +2 to −2. The numbers on the scale reflect how good or bad you think something is. Use the steps on the next page to guide you.

+2 is excellent
+1 is good
0 is neither good nor bad
−1 is not good
−2 is very poor

Activity

4 a) Create your sustainable community assessment. You will need seven columns; the first and last ones need to be bigger as you will be writing more text in these two sections. Use the example below to help.

b) Complete the three empty sections in the 'Least sustainable' column.

c) Write the location you are assessing at the top and 'Total score' at the bottom of your assessment sheet.

Most sustainable	+2	+1	0	−1	−2	Least sustainable
Active Free places to exercise, e.g. park gyms Sports facilities nearby						**Not very active** No free places to exercise No sports facilities nearby
Environmentally friendly Lots of green space Lots of recycling bins						**Environmentally unfriendly** No green space No recycling facilities
Well designed and built Lots of new buildings Uses renewable energy, e.g. solar						**Not well designed or built** Old buildings Doesn't use renewable energy
Well connected Plenty of cycle and walking routes Train stations nearby						**Poorly connected**
Thriving Many local job opportunities No empty shops nearby						**Declining**
Well served Leisure facilities nearby Healthcare services nearby						**Poorly served**

H Sustainable community assessment template

Step 1

In your local area or in and around your school, look for evidence of a sustainable community with your teacher. It is important not to rush this part and start giving each section a score before you've really explored each one. You might be surprised what you find around the corner which could make you change your mind about the score!

Step 2

Score each of the two examples for each theme on your assessment between +2 and −2. For example:

Most sustainable	+2	+1	O	−1	−2	Least sustainable
Active Free places to exercise Sports facilities nearby	✓			✓		**Not very active** No free places to exercise No sports facilities nearby

In this example, the student believes that the area has plenty of free places to exercise, giving it the highest score of +2. However, on the second option, the student thought that there were few sports facilities nearby.

Step 3

Continue down to the bottom of the sheet until each section is complete.

Step 4

Once you have completed each section, you'll need to do some quick maths. Count up both scores to give you a total. You can note these totals down on one side of your sheet.

ⓘ **Are there sports facilities in your local area?**

Activity

5 **a)** Two themes from the Egan Wheel are not included in this assessment. Which ones are they, and why do you think they are hard to measure in an assessment like this?

b) Think of one other fieldwork technique you might use to investigate the other two themes.

c) Give two examples of 'evidence of crime' you might see during your sustainable community assessment.

d) Is the data collected in these two techniques (the local sustainability survey and the sustainable community assessment) qualitative or quantitative, or a combination of the two? Explain your answer.

Creating a pictogram

A fun way of presenting your findings in geography is to create a pictogram. This uses an image to help illustrate the data you have collected. For example, to show the number of cycle racks you identified during your local sustainability survey, you could use a symbol of a bike. To show that you found five cycle racks, you could use the bike symbol five times. This makes the data quick and easy to read.

Cycle racks: 🚲 🚲 🚲 🚲 🚲

Activity

6 a) Look through the examples you provided for the local sustainability survey. Choose an icon that best represents each example.

   b) Give your pictogram a title. Using your icons, create a pictogram to show how many times you recorded each example. Here is an example.

Environmental sustainability survey   Date: 27/07/2020

Location: School grounds   Time: 2.30 pm

Evidence	Tally
Cycle racks	卌
Recycling bins	卌 IIII
Double glazing on windows	卌 卌 卌 II
Lights that turn off automatically	卌 卌 卌 II
Allotments for growing fruit, veg or herbs	II
Green roofs	I

During a fieldwork enquiry on sustainability, I discovered that our school has:

Cycle racks: 🚲 🚲 🚲 🚲 🚲

Recycling bins: ♻♻♻♻♻♻♻♻♻

Double glazing on windows: 🏠🏠🏠🏠🏠🏠🏠🏠🏠🏠🏠🏠🏠🏠🏠🏠🏠

Lights that turn off automatically: 💡💡💡💡💡💡💡💡💡💡💡💡💡💡💡💡💡

Allotments for growing fruit, veg or herbs: 🥕🥕

Green roofs: 🌱

Creating radar graphs

Radar graphs, also known as rose diagrams, are often used to show comparisons between different variables, rather like a circular bar chart. Around the chart are six variables – the themes you explored in your sustainable community assessment. Inside it you'll see the axis. For your investigation, the axis will run from +2 to −2 – the scale you used in the sustainable community assessment.

Step 1

To complete a radar graph, you will need your sustainable community assessment scores handy, and a copy of the outline of the graph. You can complete radar graphs by hand – just make sure the axis is evenly spaced out! The variables must be written out on the outside.

Step 2

Using your scores, you need to plot each result on the graph. You do this by simply putting a mark or dot with a pen onto the correct line. These marks are called data points.

Step 3

Once you have plotted all your scores onto the radar graph, it's time to connect the dots. You'll end up with a polygon, which is the unique shape made from the data you've collected. Colour in the polygon if you like.

Activity

7 Complete a radar graph for your sustainable communities assessment, using the data you collected on the previous page. If you've not had a chance to go out and collect your data, have a classroom discussion about the evidence of sustainability you've seen in and around your school. You could complete your assessment from those discussions.

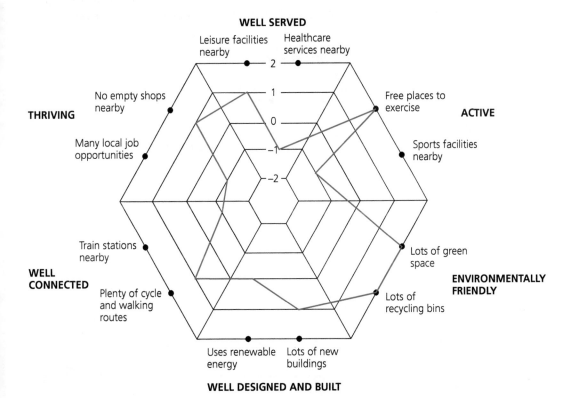

J A radar graph

Now you have presented your findings in radar graphs and pictograms, it's easier to analyse patterns and trends. The activities that follow will help you to analyse your fieldwork data.

Environmental sustainability survey

Activity

8 a) Look over the pictogram you created. Which category did you record the most frequently?

b) Which category was the least frequent?

c) Explain one reason for your answers to 8a) and b).

d) Looking at your pictogram, would you say that your local area (or the data provided in this book) is a good example of environmental sustainability or not? Explain your answer using the data you have collected and your answers above.

Sustainable community assessment

When analysing the radar graph, have a look at the polygon shape you created. Lines that move towards the centre of the chart are lower, such as −2 or −1, and lines that stretch towards the outside of the rose chart are higher (scores of +2 or +1). When explaining your findings, it is important to use the data (your evidence) wherever possible.

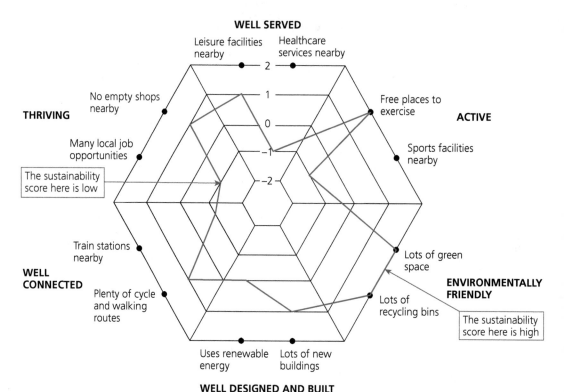

K An annotated radar graph

Activity

9 **a)** Compare the two passages below, written by two different geographers. Both of them are describing the findings from the radar graph on the previous page, but one of them is a stronger analysis than the other. Underline the parts in the second passage that make it a stronger paragraph, and explain why.

b) Write a paragraph which describes the data you have presented in your radar graph.

Geographer 1

'My radar graph was high on some scores but low on others. I thought that it was very sustainable because the area was thriving and there were plenty of local job opportunities nearby. There was also lots of green space and recycling which is very sustainable.'

Geographer 2

'My radar graph presented a range of different scores on my sustainable community assessment. In the assessment, there were many examples of ramps for wheelchairs which made the area inclusive (a score of +2), but there was some evidence of crime like graffiti on some of the fences (a score of –1). The radar graph shows high scores for green space and recycling, both receiving +2 scores, which is a good example of environmental sustainability. The lowest score recorded was –2, which was given because there was not a healthcare service nearby.'

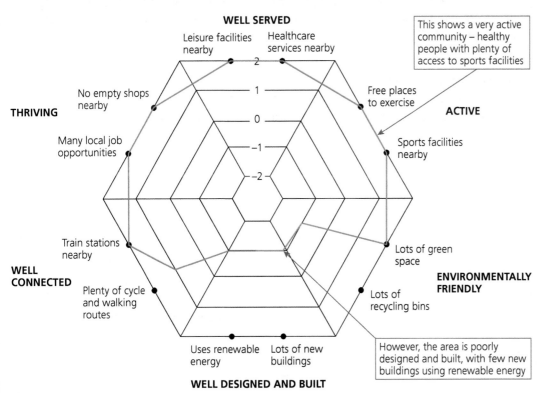

L An annotated radar graph

Evaluating your enquiry

Being able to evaluate the strengths and weaknesses of your geographical enquiry is a key part of fieldwork. In the previous enquiry, you learnt about fieldwork limitations. In addition to thinking through the limits of the fieldwork, geographers also ask the following question:

> How do you think your own thoughts and opinions could influence your findings?

In the local sustainability survey, the different examples you were looking for are very clear – recycling bins have signs on them, a light will turn off automatically if left for long enough, or an allotment might have vegetables already growing on it. Tallying up the number of times you see these examples is relatively simple. However, in your sustainable communities assessment, this isn't quite as clear-cut.

The sustainable community assessment's limitation is that the scores are very subjective. This means that it differs from person to person. If you did your fieldwork outside of the classroom, you may find it useful to see how other people scored their sustainability assessments, and compare it to yours.

The fieldwork site you chose also affects your results. For example, a residential area might have lots of recycling bins and score highly in some aspects, but there may not be much green space because of all the houses.

Activity

10 a) Which category did you find most difficult to score on the sustainable community assessment?
   Give a reason for your answer.

   b) State a limitation of the area you chose to investigate.

   c) If you were to conduct the fieldwork again, what might you do differently? Give an answer for:

   i) the fieldwork techniques you used

   ii) the area you chose for your investigation.

> My area has plenty of leisure facilities, but they are really expensive and I know people who don't use it because of the cost.

> I didn't know if my school has solar panels on the roof because I couldn't see on top of the roof!

> The buildings looked well designed and built on the outside, but they might be in a bad way on the inside.

Evaluating your data presentation

There are many ways of presenting the data you have collected. Just like the fieldwork techniques, it's important to reflect on the way you present your data as well.

Radar graphs are useful ways of presenting data as they make it really easy to spot patterns. They also allow you to compare other people's diagrams to see how they differ. For example, if you completed your own personal sustainable community assessment and were to compare your radar graph with someone else's, it would be easy to see how each one differed. The shape of your polygon might look similar, or very different!

Activities

11 a) Give one reason why a pictogram is a good way to analyse your data.

b) How else could you present your pictogram data?

12 Look at the radar graph below. Identify one or more problems with the chart.

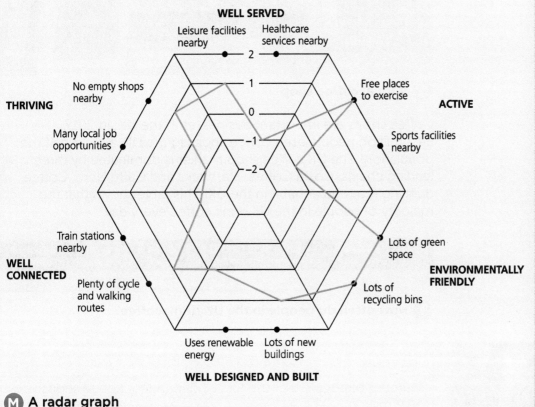

A radar graph

13 Go back to your enquiry question. Using the data you have collected, write a paragraph to explain whether you think your local area is an example of a sustainable community or not.

What sort of coffee shop would be a good business in your local area?

Learning objectives

▶ To assess the business potential of your local area.

▶ To complete a decision-making exercise and an existing business location survey.

▶ To create an annotated map and a clustered bar chart.

▶ To analyse and evaluate quantitative data.

The rise of coffee shops in urban areas

A A local coffee shop

Coffee shops are important businesses in the UK. In 2017, 34.7 million people used coffee shops in the UK, almost half the population! The UK coffee shop market is dominated by three leading chains: Costa Coffee, Starbucks and Caffè Nero. Coffee drinking is a regular habit in the UK. This table shows that the majority of people in the UK drink coffee every day.

Never	Rarely	Once a month	2–3 times a month	Once a week	2–4 times a week	Every day
11%	6%	2%	3%	5%	11%	62%

B How often do people in the UK drink coffee?

C Did you know coffee beans come from coffee cherries like these?

In the UK…

Number of people that use local independent coffee shops: 10.9 million.

Number of people that use Costa Coffee: 19 million.

Coffee shops are big business in the UK. By 2023, the number of coffee shops/cafés in the UK is expected to reach 32,230. Although leading chain coffee shops like Costa dominate high streets in the UK, there has also been a significant increase in the number of independent cafés. Other businesses, like supermarkets or fast food outlets, are also starting to open their own cafés in their shops. Even office blocks are starting to open cafés on site for their staff!

What kind of business is a coffee shop?

The economy is divided into different business sectors and includes all the types of jobs that people in the country do. These jobs are split into four employment sectors:

- The primary sector includes jobs in farming, mining and fishing.
- The secondary sector includes jobs in manufacturing and building.
- The tertiary sector includes jobs that provide services to others, such as working in coffee shops.
- The quaternary sector includes jobs in research and development, such as developing new technology.

The tertiary sector in particular has seen a huge increase: the percentage of workers in the service industry rose from 33 per cent in 1841 to 80 per cent in 2011 (the date of the most recent Census when this book was written). In London, 91 per cent of the city's economy is in the service sector – the highest in the UK.

Activity

1. a) Create a bar chart to present the data in the table on page 24.

   b) Think of the work that goes into making a cup of coffee, from bean to cup. Using the definitions of primary and secondary sector, fill in the following table with one job you might have in each sector to make everything that goes into a cup of coffee.

   c) In the previous chapter, you learnt about sustainability. Given that there are more and more people using coffee shops, describe three ways you could make a coffee shop more sustainable.

Primary	Secondary	Tertiary
		Baristas to serve the coffee in a coffee shop.

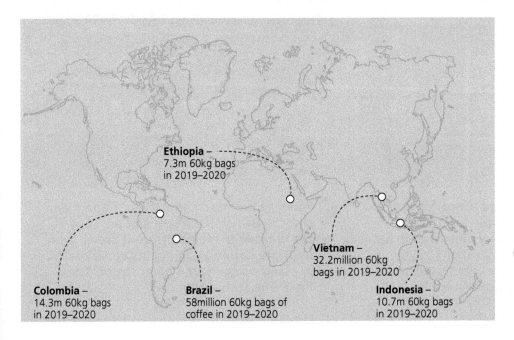

Ethiopia –
7.3m 60kg bags
in 2019–2020

Vietnam –
32.2million 60kg
bags in 2019–2020

Colombia –
14.3m 60kg bags
in 2019–2020

Brazil –
58million 60kg bags of
coffee in 2019–2020

Indonesia –
10.7m 60kg bags
in 2019–2020

D Here are some of the places around the world that produce a huge amount of coffee each year

Collecting data

What sort of coffee shop would thrive in my local area?

In the area where you live, you may have noticed different kinds of coffee shops. In this enquiry, you will imagine yourselves as business owners and consider opening up a new café or coffee shop. There are four types of café you are considering. These are:

- an independent coffee shop
- a traditional British café
- a pay-what-you-can coffee shop
- a popular coffee shop chain.

The four businesses you could choose from all have their own target markets and unique selling points. An area that has a lot of tourists, like a major city, might find that a popular coffee shop chain will thrive as people are familiar with their menu, whereas an independent coffee shop stocking expensive coffee beans might not succeed in more deprived urban areas.

E Independent coffee shops have unique features. They may offer specialist coffee beans, or be more experimental in what they offer. Independent coffee shops sometimes raise their money from crowdfunding – an online donation scheme for businesses.

F Traditional British cafés offer affordable food and coffee. They're well-known for their fried breakfasts and tend to be open early for workers to eat before starting their day. They tend to be family-run and often have a community feel.

G A pay-what-you-can café is not-for-profit. Rather than charge people a set price, customers put their money into a discreet 'contributions' box. They usually have a small, simple menu and are popular with low-income families or people experiencing homelessness.

H A popular coffee shop chain is a familiar sight in British towns and cities. They are located in many different areas, and multiple coffee shop chains can often be spotted on the same main street. They have one standard menu.

In order to narrow down which coffee shop/café would thrive in your local area, you can use a business decision-making exercise. In this exercise, geographers list the positives and the negatives (also known as pros and cons) of each type of business.

Fieldwork technique 1: a business decision-making exercise

A business decision-making exercise requires a table listing all the possibilities for your business, and a list of pros and cons.

To complete the exercise, you will need to either explore your local area as a school group, or explore it virtually through Google Street View. While exploring, list the pros and cons of your local area and each of the possible business options. For example:

Type of business		My local area: West Ham	Best choice?
Independent coffee shop	Pros	There are only two other cafés in this area	𝑥
	Cons	There's a big Starbucks on same street – people might choose the chain café instead	
Traditional British café	Pros		
	Cons		
Pay-what-you-can café	Pros		
	Cons		
Popular coffee shop chain	Pros		
	Cons		

Activity

2 **a)** Complete the business decision-making exercise for your local area. List at least one pro and one con for each type of coffee shop.

b) Using the answers from your business decision-making exercise, which of the four coffee shops did you decide would be the most successful in your local area?

Identifying the best location for a new coffee shop

Now you have selected what kind of coffee shop you want to open, the next step is to consider the exact location for it. This includes thinking carefully about a number of factors, including:

- Premises: what is the area like?
- Transport: can customers easily access the location?
- Market: do I have competition from other businesses?

Fieldwork technique 2: a business location survey

You can use a business location survey to examine different potential areas for a new business. In this business location survey, you will examine three different potential areas for your coffee shop.

A business location survey is a quantitative fieldwork technique. It requires you to score each location based on three key factors – the premises, transport and the market – on a score of 1–5. Like other scales, 1 is a poor score, while 5 is an excellent score. To compare each location, you can count the total score and write it at the bottom. You may wish to take photos as evidence.

Activity

3 a) Using Google Street View and/or a group discussion, propose three suitable locations for your coffee shop. There are many different locations that might be suitable, for example a high street location, near a train station or beside a busy road.

b) Explain why you have chosen these three locations for your chosen coffee shop.

c) Now you have identified the three locations to survey, create a business location survey using the example below.

For each location, give a score for the factors below, from 1 to 5, where 1 is poor and 5 is excellent. Add the scores to work out a total for each location.		Location 1	Location 2	Location 3
Premises	• suitable premises – either existing building or new site • available premises – either to buy or rent, new or empty • thriving local high street	• • •	• • •	• • •
Transport	• accessible public transport • car parking provision • easy cycle and/or pedestrian access	• • •	• • •	• • •
Market	• footfall – number of potential customers passing by • types of people – relative to target group, in terms of age and/or income • no similar businesses nearby	• • •	• • •	• • •

The map below shows the many boroughs of London.

Forest Gate is located in the Borough of Newham, in East London. Forest Lane is a road that runs through Forest Gate. On Forest Lane you will find:

Two small shops for sale.

A new commercial building opening up close to the train station.

A busy high street.

A small number of other cafés nearby.

Numerous bus routes and a busy train station.

High footfall, with lots of young people and parents with prams.

No car park, unless you stop in the local supermarket and pay for parking.

Good pedestrian access and some cycle lanes.

FOREST LANE

Activity

3 d) Give one reason why shops with a 'for sale' sign might not be positive news for your business.
 e) How could you more accurately measure 'footfall' in each location?
 f) What is the target group for your coffee shop? Explain one reason why.

Presenting your data

To present the qualitative data from the business decision-making exercise, you can create an annotated map. To present the quantitative business location data, you can create a clustered bar chart.

Making an annotated map

To present the pros and cons of each location, you can annotate a map to show each location clearly and summarise your findings.

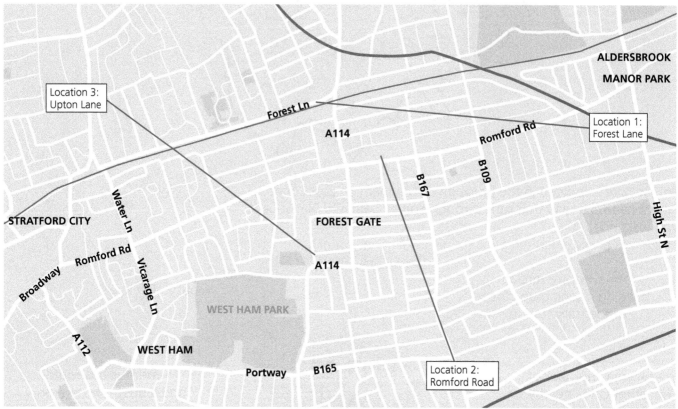

ⓘ **Forest Gate, E7, London**

Step 1

Print out a map of the local area you are using for your future coffee shop. Ensure you have all three possible locations clearly marked out.

Step 2

Stick the map into the middle of a large piece of paper, giving you enough space to annotate around the outside.

Step 3

Write the pros and cons of each location around the outside. Draw a straight line from the annotation to the location. It's important to use a straight line rather than an arrow on annotated maps because arrows tend to show directions or flows in geography (for example, the direction wind or water is travelling).

Step 4

If you took photographs as evidence during your fieldwork, print one or two photos per location and stick them onto the map near your annotations.

Location 3: Upton Lane
Pros: Close to a secondary school – parents might use it
Cons: Similar business on nearby Romford Road

Location 1: Forest Lane
Pros: Excellent transport links
Cons: Independent café nearby seems extremely popular – competition may be strong

Location 2: Romford Road
Pros: High footfall
Cons: Big supermarket café on same road

J **Annotated map of Forest Gate**

Presenting your data

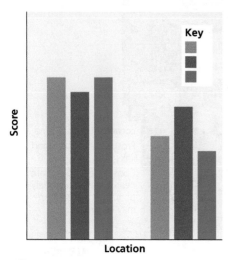

Score

Location

K **Example of what a clustered bar chart looks like**

Key

Clustered bar charts

The data collected in the business location survey included scoring (1–5) the premises, transport and market potential for your new business. As this data is quantitative, it can easily be presented in a range of charts and graphs, including a clustered bar chart.

A clustered bar chart is a chart with more than one bar, or column, of data next to one another. It's useful to show how the premises, transport and market in each location differ from one another.

Step 1

To begin your clustered bar chart, give your chart a title and create the y-axis (vertical) and x-axis (horizontal). The y-axis will be your business location score, and your x-axis will be the location. The highest number you will need the y-axis to go up to will be 15, as the highest possible score on the survey is 5, and there are three subcategories. The x-axis will have your three locations. For each location, there will be a 'cluster' of three bars – one for premises, one for transport and one for market.

Step 2

Start with the first location. The first bar will show your Premises total score. Find the total score by adding the three scores for Premises. Create the first bar for Premises.

Forest Lane: Premises: 4 + 2 + 5 = 11

For each location, give a score for the factors below, from 1 to 5, where 1 is poor and 5 is excellent. Add the scores to work out a total for each location.	Forest Lane	Upton Lane	Romford Road
Premises: • suitable premises – either existing building or new site • available premises – either to buy or rent, new or empty • thriving local high street	• 4 • 2 • 5	• 2 • 2 • 3	• 4 • 2 • 5

Step 3

Move onto the next category, Transport. Count the total score and create a bar chart with a small space between each bar. For example:

Forest Lane: Transport: 5 + 1 + 4 = ☐

For each location, give a score for the factors below, from 1 to 5, where 1 is poor and 5 is excellent. Add the scores to work out a total for each location.	Forest Lane	Upton Lane	Romford Road
Transport: • accessible public transport • car parking provision • easy cycle and/or pedestrian access	• 5 • 1 • 4	• 3 • 2 • 4	• 5 • 1 • 4

Step 4

Continue with the final category, Market:

Forest Lane: Market: [＿＿＿＿＿] = [＿＿＿]

For each location, give a score for the factors below, from 1 to 5, where 1 is poor and 5 is excellent. Add the scores to work out a total for each location.		Forest Lane	Upton Lane	Romford Road
Market	• footfall – number of potential customers passing by • types of people – relative to target group, in terms of age and/or income • no similar businesses nearby	• 3 • 5 • 3	• 2 • 3 • 1	• 5 • 5 • 1

Step 5

Add a key so that it is clear what each bar represents.

Step 6

Continue creating cluster bar charts for the next two locations on the same piece of graph paper. Ensure you have space between each 'cluster'!

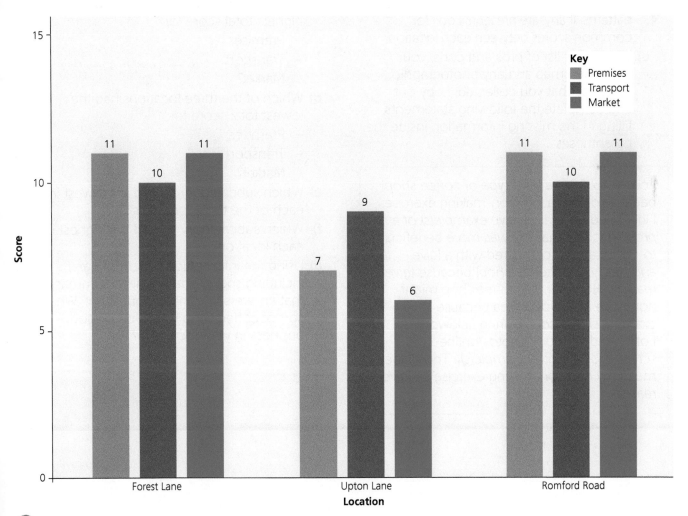

L **A clustered bar chart**

Analysing your business decision-making exercise

Now that you've presented your data, it's time to take a close look at it by analysing and evaluating your fieldwork.

Analysing your business location survey

When analysing the results, you may notice some common patterns. For example, you may have spotted that there are more low scores for 'car parking provision' under the transport section of your business location survey, compared with the other two transport variables, or subcategories. These patterns are important to spot in data analysis, to show if there are any weaknesses to your potential business location.

Activity

4 a) Using your business location survey, what patterns, if any, are present? Look for common scores between each location.

b) Using your list of pros and cons, your annotated map and any photographic evidence that you collected, copy out and complete the following statements, filling in the missing information inside the parentheses:

I chose to open a [*give type of coffee shop*] because during a decision-making exercise, I identified that it had [*give example(s) of a pro*]. I decided that this was more beneficial to my local area compared with a [*give another type of coffee shop*] because [*give reason*]. However, my coffee shop might not thrive in my local area because [*give example(s) of a con*]. During fieldwork, I observed and took photos/online screenshots of [*give example(s)*]. This helped me in my decision-making exercise by [*give reason*].

c) Which of the three locations had the highest total score for:
  - Premises
  - Transport
  - Market?

d) Which of the three locations had the lowest total score for:
  - Premises
  - Transport
  - Market?

e) Which subcategory scored the lowest in each of the three locations?

f) Which subcategory scored the highest in each location?

g) Using the information in activity 4a)–f), including specific data, write a summary that answers the enquiry question: 'What sort of coffee shop would be a good business in your local area?'

Evaluating your enquiry

Use the following questions to evaluate the data you have collected
and presented in this enquiry.

Activity

5 a) Look back at your business decision-making exercise. How did you come to a final decision on
what type of coffee shop to open?

b) State one limitation you had with annotating your map.

c) Describe one issue with only including the total score of the business location survey in the
clustered bar chart.

d) Using your answer to 5c), explain how you might overcome this issue.

e) What other kind of data presentation technique could you use to show the results of your
business location survey?

f) Choose another category you could add to your business location survey.

g) Read the following extracts. Do you agree with what the student has said about their fieldwork
evaluation? Explain why for each extract.

> In my business location survey, I used quantitative data to give each
> area a score between 1 and 5. As I used numbers, it was more
> accurate and therefore was not subjective. When I could not decide on
> a score, I gave it a 3 because that was basically in the middle of 1–5.

> Romford Road was over a mile and a half long so we
> chose to do our fieldwork on a designated one-mile
> part of the road. I clearly marked this area on my map.
> Although we had a clear area to work in on our fieldwork,
> it meant that we couldn't collect data on the full site and
> may have missed key examples of data.

> My clustered bar chart clearly
> showed that my coffee shop would
> thrive in my chosen location, with
> scores of 3, 4 and 4.

4.1 What are the different microclimates around my school?

Learning objectives

▶ To understand the difference between weather and climate.

▶ To measure heat and wind around my school.

▶ To understand how physical features affect microclimates.

▶ To create a scatter graph of my findings.

▶ To analyse and evaluate my enquiry findings.

What is a microclimate?

Every day, we are affected by the weather. Weather is the state of the atmosphere at a particular place and time. Climate, however, is the average weather over a longer period of time. A microclimate is the local climate of a small-scale area, such as a garden or part of a city.

9 a.m., 30 April 2019, Geo School, North London	30 April 2015–30 April 2020, UK	30 April, outside Science Block at Geo School
Weather	Climate	Microclimate

Depending on where you live in the UK, there will be very different microclimates. Upland areas tend to be cooler than lowland areas. In fact, the temperature can fall 5–10 degrees Celsius per 1000 m ascent. Forests also tend to be cooler, thanks to trees shading the ground beneath them.

Urban areas are particularly warm compared to upland areas and forests. Urban areas tend to be warm due to a process known as the urban heat island effect. Heat islands are formed for a number of reasons:

● Buildings and concrete store and release heat during the day and night.

● Urban pollution, including smog (a pollution layer), traps radiation and keeps areas warm.

● Light and heat are reflected from buildings, particularly skyscrapers.

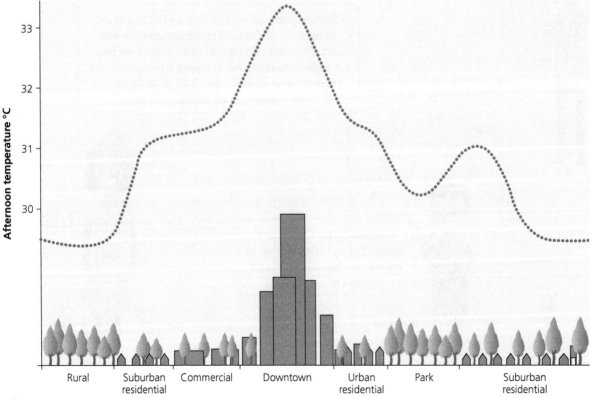

A A microclimate graph

Investigating microclimates

In this investigation, you will complete an enquiry into the different microclimates in and around your school. Investigating microclimates might involve measuring any of the following variables:

- temperature
- wind direction
- wind speed
- light
- precipitation.

To measure these, you need some fieldwork equipment. These are some of the instruments you might use in your investigation:

- anemometer
- light meter
- thermometer
- compass
- rain gauge.

There are many factors which influence local temperature and wind, and therefore affect microclimates.

- Physical features, such as hedges, provide shelter from the wind, which means the area you study might have a different microclimate from an area without physical shelter.
- Individual buildings can make areas warmer as they give off heat, but buildings close together can also make microclimates very windy. The wind tunnel effect is the name given to fast-travelling wind rushing between two buildings close together.
- The direction that a place is facing can also affect the microclimate of an area. A building facing the sun, for example, will be warmer than one in the shade. The direction a place is facing is called aspect.

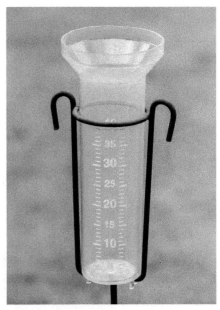

B A rain gauge

Activity

1 a) Match the microclimate variables (above) to the instruments you might use in your investigation.
   b) Name two examples of types of precipitation.
   c) State the unit that temperature is usually measured in.
   d) State two physical features, other than hedges, that could provide shelter from the wind.
   e) Explain one way that a hedge might affect wind speed and direction.

Site A: sports field	Site B: entrance	Site C: school gardens	Site D: car park

C Examples of possible school data collection sites.
Other examples of sites could include tennis courts,
science block and swimming pool

Preparing for your enquiry

Before you begin your enquiry, you will
need to select four different sites. You may
want to have a discussion in class about
which sites you'd like to choose.

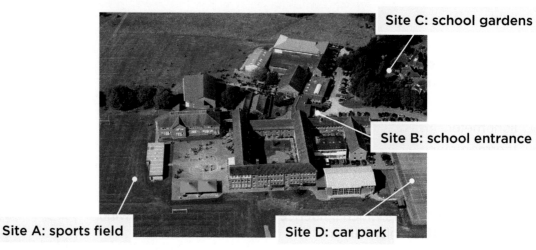

Site C: school gardens

Site B: school entrance

Site A: sports field

Site D: car park

D Annotated aerial view of school grounds

Activity

2 a) Print out a satellite
image map of your
school and stick it
onto a plain piece of
A3 paper, with space
around the image
to write in. Annotate
the map to show
how the buildings,
aspect and physical
environment of the
school could affect
the microclimate.

You can use Google Maps, or another GIS program to look at your
school from a satellite view and create your own annotated map.
Type your school name into the search bar and select 'satellite
view' at the bottom. A direct bird's eye view might make it easier to
choose your sites, as you can see the buildings' aspect, the shelter
provided by physical features, and any open spaces.

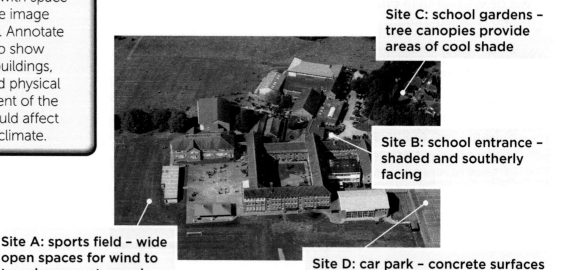

Site C: school gardens –
tree canopies provide
areas of cool shade

Site B: school entrance –
shaded and southerly
facing

Site A: sports field – wide
open spaces for wind to
travel across at speed

Site D: car park – concrete surfaces
absorb heat, while nearby hedges
provide shelter from the wind

E Annotated image of microclimates around school grounds

What you will be measuring

For this enquiry, you will be measuring three variables in each area: wind (both speed and direction), temperature and cloud cover.

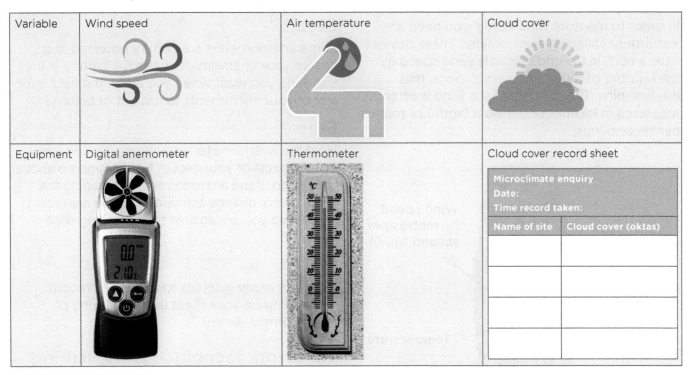

Variable	Wind speed	Air temperature	Cloud cover
Equipment	Digital anemometer	Thermometer	Cloud cover record sheet

Before you go out and collect your data, you need to create your recording sheet.

Activity

2 b) Create a recording sheet for your data collection, using the template below as a guide:

Microclimate enquiry
Date:
Time record taken:

Name of site	Wind speed (km/h)	Air temperature (°C)	Cloud cover (oktas)
Site A: sports field			
Site B: entrance			
Site C: school gardens			
Site D: car park			

F **Microclimate data collection template**

Fieldwork technique 1: measuring wind speed

In order to measure wind speed, you need an instrument called an anemometer. These devices have a built-in fan and calculate wind speed by the number of rotations, or revolutions, that the fan spins. The strength of the wind is often measured in kilometres per hour (km/h) or metres per second (m/s).

Wind wheel

Wind speed (in metres per second (m/s))

Temperature

G A digital anemometer

To take a wind speed reading you need to follow these steps:

Step 1

Some anemometers are battery powered. If so, ensure your anemometer has a full battery – the last thing you want when you are in the field is for any of your instruments to run out of battery!

Step 2

Hold the anemometer away from you and away from the rest of your group. Hold it high up above your head. If the anemometer isn't reading the wind speed, change the direction you are facing slightly, so you are against the prevailing wind direction.

Step 3

Once the reading settles to a number, record the reading on your sheet using your unit of measurement (km/h).

Fieldwork technique 2: measuring air temperature

The temperature of each site is measured using a thermometer. There are two types of thermometer you might use in your investigation:

- mercury thermometer
- digital thermometer.

H Mercury thermometer

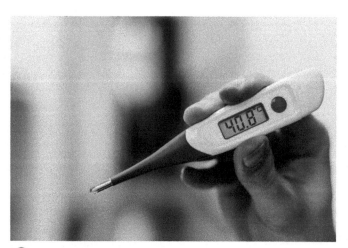

I Digital thermometer

To measure air temperature, hold the thermometer away from your body and away from direct sunlight. Wait a little while for the reading to settle, then record the score on your sheet.

Fieldwork technique 3: recording cloud cover

Cloud cover is a measure of the amount of sky obscured by clouds. It is measured by a unit called oktas. Oktas help us to understand how much of the sky is covered by cloud. Each okta represents one-eighth of the cloud cover.

Cloud cover oktas are shown by using the following symbols:

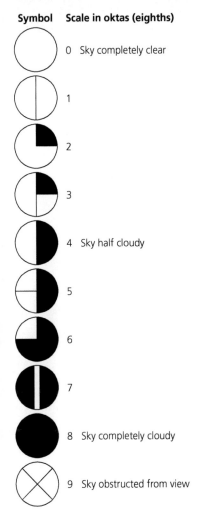

Symbol	Scale in oktas (eighths)
	0 Sky completely clear
	1
	2
	3
	4 Sky half cloudy
	5
	6
	7
	8 Sky completely cloudy
	9 Sky obstructed from view

J Cloud cover oktas

To record your cloud cover data:

Step 1
Look up at the sky. Using your opinion, assess how covered by clouds you think the sky is.

Step 2
Record this data by drawing a circle with the correct symbol, using the key in Figure J, of cloud cover.

K Blue sky with some cloud cover

Activity

3 **a)** Explain why you have chosen your sites as good locations for your investigations.

b) Create a microclimate data collection template, adapting the one provided on page 39.

c) Give one reason why you should hold the anemometer away from your body and above your head when recording wind speed.

d) Sailing is a sport where you might use an anemometer to help you better understand the wind conditions. Can you think of another example?

e) How many oktas of cloud cover do you think are shown in Figure K? Draw the answer using the okta scale symbols.

Presenting your data

You can use a scatter graph and your annotated map to present your data.

You may use the data you have collected from your school, or the data from the table below.

Microclimate enquiry Date: 27 August 2019 Time record taken: 9.15 a.m.			
Name of site	Wind speed (km/h)	Air temperature (ºC)	Cloud cover (oktas)
Site A: sports field	8	21.0	1
Site B: entrance	6	20.0	2
Site C: school gardens	3	20.5	4
Site D: car park	2	23.0	1

L Microclimate enquiry: collected data

Annotated map

An annotated map that includes the data you have collected is a great way to present your results.

Step 1

Print a new map of your school or local area and stick it onto plain A3 paper. Locate and label each of the sites you chose for your enquiry. Use the example provided on page 38 to help you.

Step 2

Annotate the map with the cloud cover oktas for each site.

Step 3

Add the data for air temperature underneath the okta. Ensure you use the correct unit of measurement for temperature.

Step 4

Finally, write the wind speed underneath – with the correct unit of measurement – for each site.

Scatter graph

In order to show the different data across the different sites, you can create a scatter graph. Scatter graphs are one way to show the relationship between different sets of data, such as temperature and wind speed.

Step 1

Plot your x-axis. This is the axis that will show the wind speed data, measured in km/h.

Step 2

Plot your y-axis, which will be where you put your temperatures. Have a look at how much or how little the temperature goes up or down each site. This will be important when you plot your scores.

Step 3

Plot the data for each site. For example, if Site A has a wind speed of 2 and a temperature of 21, go along the x-axis to find the wind speed, then up to find the temperature and make a cross or dot.

Step 4

Label each of the data points with the name of the site.

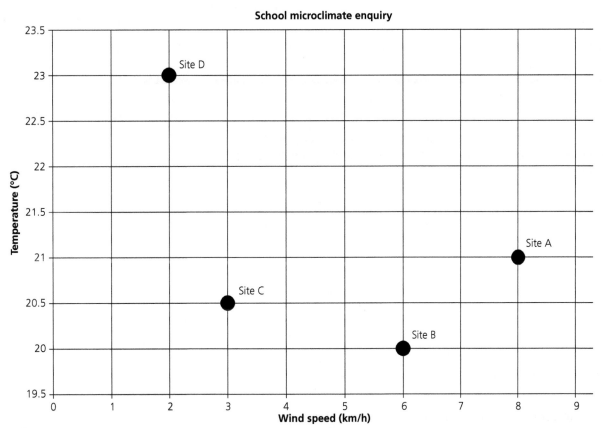

Ⓜ **Scatter graph for a school microclimate enquiry**

Analysing your data

So far in this enquiry, you have explored how temperature, wind speed and cloud cover affect microclimate. When analysing microclimates, it is important to think about the whole geography of each site – for example, whether there are lots of buildings, open space, trees or shade. Looking through your data, or the data provided in this chapter, answer the following questions:

Activities

4 **a)** Which site had the highest temperature? Explain one possible reason.
   b) Which site had the highest wind speed? Explain one possible reason.
   c) Choose two sites. How is the microclimate of each site affected by nearby buildings?
   d) Which site would be the best place to put a solar panel? Explain your answer.
   e) Which of the locations would be the best place to put a wind turbine? Explain your answer.
   f) How might the data change if you recorded it during the night?
   g) How did cloud cover affect each result?
   h) How did the wind speed affect the temperature of each site?
   i) Was there any evidence in your data of 'urban heat island effect'? If so, give examples from each site to explain why.

5 Using the three steps below, and some of the answers to 4a)–i), write two paragraphs analysing your data.

Writing up your analysis

In geographical enquiries, being able to write up your analysis is an important skill. A good pattern to use to help you write your analysis might look like this:

Describe	Examples	Anomalies
My school's microclimate varied depending on location. This may be because…	The warmest microclimate was the car park, at 23 degrees.	We recorded one unusual result.

Step 1

Start by describing the overall microclimate of the area. Which sites did you visit, and why did you pick them?

Step 2

Use some examples and draw on specific data. For example, 'At Site E, we recorded our highest temperate – 23 degrees Celsius.'

Step 3

Try to identify any anomalies. Where possible, think of a reason why these are anomalies – what might have caused these unusual results?

Anomalies are differences in the data which you might not expect. For example, if there is a pattern or trend in your graph, but one or two results do not seem to fit into that trend, these could be described as anomalies.

Evaluating your enquiry

There are benefits and limitations to all kinds of fieldwork enquiries. Using instruments in data collection, like a thermometer or anemometer, is no different.

When evaluating fieldwork data, it is useful to think about the possible sources of error that might have affected your results. Here is one example:

Equipment	Anemometer
Measures	Wind speed
How data was measured and recorded	We held up the instrument and waited for a reading to settle. The data was recorded on a recording sheet.
Sources of error	When holding up the anemometer to get results, the score on the screen changed depending on how high we held the anemometer. Sometimes, wind speed was recorded from different heights by accident, because we all took it in turns to hold the instrument up.

Sources of error are common in many fieldwork enquiries. One way to combat the sources of error above is to ensure that the same person takes the reading for wind speed in each location. One other option is to have that same person take more than one reading at each site. For instance, taking three readings of wind speed in a five-minute period and calculating the mean (average) might be one way.

Activity

6 a) Create a table like the one below. List up to three sources of error for each of the following:

Equipment	Thermometer
Measures	
How data was measured and recorded	
Sources of error	

Equipment	Okta recording
Measures	
How data was measured and recorded	
Sources of error	

b) Do you think you chose the right sites for your enquiry? Explain your answer.

c) Give one reason why measuring cloud cover over the course of one hour may not have been the most suitable fieldwork technique.

d) How might you use the following objects in a future microclimate enquiry?
   - A wind sock could measure...
   - A bucket could collect...to measure...
   - A cup of water could help measure...

e) Give one limitation of each of these fieldwork instruments.

Learning objectives

▶ To understand why we need to protect our coastline.

▶ To measure different types of coastal protection and calculate their cost.

▶ To measure the movement of material along a beach and present the data.

▶ To use secondary data to estimate property values in a seaside town.

▶ To work out if it is worth protecting the beach in a seaside town.

Why are coastlines important and why do they need to be protected?

The British Isles is a place that has been shaped by the sea. It is a dynamic and changing environment that includes approximately 16,000 km of coastline (depending on how you measure it). The coastline is the boundary where the land meets the sea.

All around the British coast there are hundreds of seaside towns and villages. The seaside is popular with tourists who go there for their holidays to fish, swim, sail or maybe just to sunbathe on the beach. Of course, there are some lucky people who live at the seaside and can do those things all year round!

However, seaside towns are not without their problems. The sea has the power to change the shape of the land by means of erosion, transport and deposition. In storm conditions, powerful waves erode the coastline, wearing away the rock and destroying beaches. In a severe storm, towns can be flooded. One way to prevent erosion or flooding is to build a sea wall along the coast, behind the beach, to stop waves reaching the land (Figure A).

Even under normal conditions, waves transport sand, shingle and pebbles along the beach. This process is called longshore drift. Over time, unless the material is replaced, the beach will disappear. Without the beach to absorb their energy, it is easier for waves to erode the land. One way to prevent that happening is to build groynes – barriers built at right angles to the beach to prevent the movement of material (Figure B).

A A sea wall built behind the beach to protect the land

B Groynes built to prevent material moving along the beach

Activity

1 Look closely at Figures A and B.

a) Explain how sea walls and groynes can help to protect the coast. Write a sentence about each one.

b) Do you think sea walls and groynes are a good thing to have on a beach or not? Think about their appearance and what their job is. Explain your point of view.

Herne Bay – a seaside town

Herne Bay is a seaside town in north Kent on the North Sea coast. It is about 80 km from London, where the Thames estuary widens to meet the sea. The town has a beach stretching from east to west, attracting many visitors from London.

When the wind is blowing from the north-east, Herne Bay is exposed to the full power of the waves from the North Sea. In February 1953, during a particularly bad storm that affected the entire east coast of England, Herne Bay was flooded (Figure D). The flooding caused £250,000 worth of damage, which would be millions of pounds in today's money. Fortunately, there were no deaths in Herne Bay due to the floods, but over 300 people drowned along the rest of the east coast.

Since 1953 coastal protection in Herne Bay has been improved to reduce the danger of flooding. But with sea levels rising around the world, and storms becoming more frequent, there is no guarantee that there won't be more floods in future.

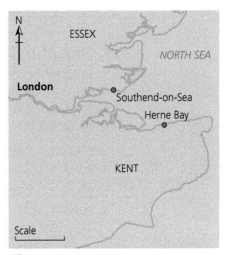

C **The location of Herne Bay in Kent**

D **Flooding in Herne Bay in 1953**

Activity

2 **a)** Why does Herne Bay's location mean that it is affected by winds blowing from the north-east?

   b) What is your nearest seaside town? Which coast is it on? Use an atlas map or Google Maps to find out.

   c) Carry out some research about Herne Bay, or your nearest seaside town, to find out about any storms that have occurred there since 1953. When did they happen and how did they affect the town?

Collecting data

Coastal protection

Your teacher may plan a visit to a seaside town where you will be able to investigate coastal protection for yourself. This chapter will help you to prepare for your visit. If no visit is planned, you can use the data from Herne Bay in this chapter to carry out your own virtual investigation.

Remember the question that you have to investigate: 'Is it worth protecting the coast in a seaside town?' You will use three data collection methods to help you answer this enquiry question:

- Identify and measure different types of coastal protection in order to calculate their cost.
- Measure the movement of material along the beach as a result of longshore drift to find out if protection is needed.
- Map and count the number of seafront properties and estimate their total value with the help of secondary data.

Types of coastal protection

We have already come across two types of coastal protection – sea walls and groynes. The table below mentions two more and gives details of how each type of protection works and how much it costs.

Type of coastal protection		Details and costs	Advantages and disadvantages
	Sea wall	A concrete barrier built behind a beach to protect the land. It often has a curved surface to deflect the power of the waves back into the sea. Sea walls cost £10,000 per metre to build.	• The top of the sea wall can provide a promenade where people can walk. • A sea wall restricts access to the beach and can spoil the natural look of the beach.
	Rock armour	Large rock boulders usually placed at the back of the beach. The rocks absorb the power of the waves in order to protect the land. Rock armour costs £2,000 per metre to build.	• Rock armour can be used in places where it would be difficult to build a sea wall. • Rock armour is ugly and difficult for people to climb over on the beach.
	Groyne	Wooden or concrete barriers built at right angles to the beach to prevent the movement of sand, shingle or pebbles along the beach. Groynes cost £5,000 each and are built about 20 metres apart.	• Groynes act as windbreaks where people can shelter or even fish from. • Groynes make it difficult to walk along the beach and are hazardous to climb over.
	Beach replenishment	Sand, or other beach material, is imported and spread over the beach to replace the material that has disappeared. Beach replenishment costs £100 per metre.	• Beach replenishment creates what looks like a natural beach. • Beach replenishment needs to be repeated as waves continually remove the sand.

E Types of coastal protection

Fieldwork technique 1: identify and measure coastal protection

Step 1

Study the table describing the different types of coastal protection (Figure E) so that you will be able to identify them when you see them.

Step 2

Using a map, walk along a stretch of coastline to identify the types of protection. Mark what you find onto a map like Figure F and choose different colours or symbols to mark each type of protection. Try to be as accurate as you can. For example, if you are showing a sea wall or rock armour, mark where it begins and ends and draw a line to show how far it stretches along the coast. With groynes, draw each groyne at right angles to the coast. It is unlikely that you will actually see the beach being replenished on the day you go, so you won't be able to show that.

Step 3

When you have completed the map, use the scale to work out the length of each type of coastal protection. In the case of groynes, count the total number. From this you can calculate the cost of each type of protection to work out the total cost.

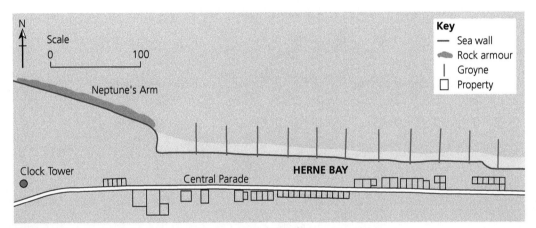

F Map showing types of coastal protection at Herne Bay

Activities

3 Look at Figure F. Identify the types of coastal protection on the map of Herne Bay – measure them using the scale and count the groynes. Then, write your lengths and numbers in the second column of the table below.

4 Complete the rest of the table below.
   a) Write the cost per metre (or per groyne) of each type of protection in the third column, using the costs in Figure E.

b) Calculate the cost of each type of protection by multiplying the length (or number) by the cost per metre (or per groyne) – you worked these out in Step 3. Write the costs in the fourth column.

c) Add up these costs to work out the total cost of coastal protection along this stretch of coast in Herne Bay.

Type of coastal protection	Length of coastal protection (or number of groynes)	Cost per metre (or per groyne)	Cost of each type of coastal protection
Sea wall			
Rock armour			
Groynes			
Total cost			

Measuring longshore drift

Herne Bay's location on the north Kent coast means that north-easterly winds across the North Sea cause longshore drift to happen from east to west. On other coastlines the direction of the wind may be different, causing longshore drift to happen in other directions. Longshore drift is important to understand in seaside towns because it can cause the beach to disappear which could make it easier for waves to erode the land behind the beach. The disappearance of the beaches could also affect tourism.

Waves are driven by the wind, so they hit the coast at an angle and create a swash that moves material up the beach. When the waves break, the backwash rolls straight back down the beach, dragging material with it. Repeated waves cause material to be transported along the beach in a zigzag (Figure G).

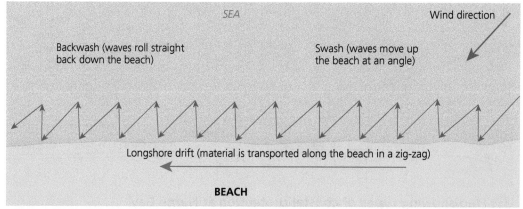

G **Longshore drift transports material along the beach in a zigzag**

There are two ways to measure the movement of material along a beach as a result of longshore drift. One way is to measure the movement on the day that you visit. The other way is to measure the effect of this movement over a long period of time.

Fieldwork technique 2a: measure longshore drift on the day

You will need an orange (or object of no value that floats on water), a tape measure, a metre rule (or long stick), a phone (or watch) to keep the time and a compass. Work in a small group.

Step 1

First, try to find out the wind direction on the day that you visit the seaside. You can do this by licking your finger and holding it up in the air to see which side of your finger feels colder – that is the direction the wind is coming from. You may find that you need a compass to work out the exact direction though. You would expect longshore drift to happen in the same direction that the wind is coming from.

Step 2

Lay your tape measure along the edge of the beach, near to the sea. If there are groynes try to lay the tape measure away from the groynes because they stop longshore drift. Lay the tape measure out for 20 metres and stand at the 10-metre point, half way along the tape measure.

While carrying out this step be careful not to get too close to the water's edge.

Step 3

Start your stopwatch (either on your watch or on your phone). Throw the orange into the sea, as far as you can throw, at right angles to the beach. Then record the movement of the orange to the east or west along the beach every 10 seconds. See Figure H for a visual guide on performing this task. The best way to do this is for you to move along the tape measure in the same direction as the orange moves, holding up the metre rule to make sure you are looking in a straight line, and record how far the orange has moved from the distance on the tape measure. Keep going for five minutes, or until the orange reaches the end of the tape measure.

Step 4

Repeat the method in case the measurement you got the first time was unusual. You can then take an average of the two measurements – you would expect them to be similar unless the wind changes. The direction of movement should be the same as the wind direction. The stronger the wind, the bigger the waves and the further the orange will move.

Activity

5 Study the movement of the orange along a beach in Figure H. Record the movement in the table below. The first movement is done for you.

Time (minutes + seconds)	Reading on tape measure (metres)	Distance moved by orange (metres + direction)
0.00	10.00	0.00
0.10	09.00	1.00 west
0.20		
0.30		
0.40		
0.50		
1.00		

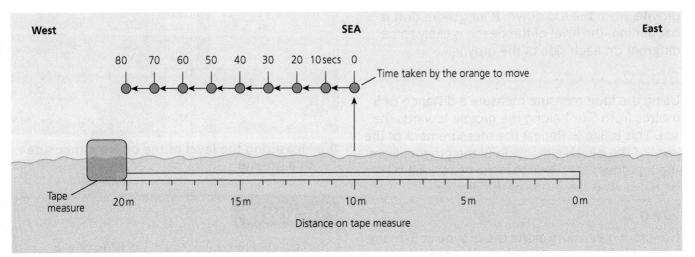

H The movement of an orange along a beach

Groynes help to prevent longshore drift. As waves transport material along the beach, it is likely to build up on one side of the groyne where the material is trapped. This shows the direction that the longshore drift is coming from. On the other side of the groyne, longshore drift transports material further along the beach to the next groyne. Over many years, material builds up on one side of the groynes all the way along the beach. This shows longshore drift happening on a beach over a long time and an example of this can be seen in Figure I.

I Sand builds up on one side of a groyne because of longshore drift

Fieldwork technique 2b: measure longshore drift over a long period

You will need to find a groyne on the beach. You will also need a tape measure and a metre rule. Work in a small group.

Step 1

Start at the end of the groyne close to the back of the beach. This is Site 1. Use the metre rule to measure the level of the beach either side of the groyne, from the top down. If longshore drift is happening, the level of the beach is likely to be different on each side of the groyne.

Step 2

Using the tape measure measure a distance of 5 metres from Site 1 along the groyne towards the sea. This is Site 2. Repeat the measurement of the level of the beach either side of the groyne, from the top down. It is likely that the beach will still be higher on the same side of the groyne.

Step 3

Continue measuring along the groyne at 5-metre intervals until you reach the sea. Don't take any further measurements because it is dangerous to go into the sea.

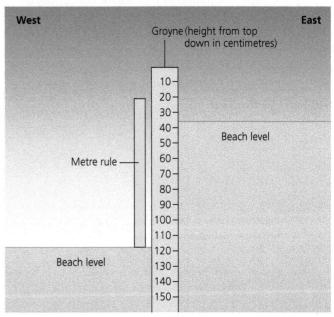

J Measuring the level of the beach either side of a groyne

Activity

6 Look at Figure J. What is the level of the beach from the top of the groyne:
a) east of the groyne?
b) west of the groyne?
c) What does this tell you about the direction of longshore drift on the beach? Explain your answer.

Fieldwork technique 3: map and count the properties close to the beach

Properties along the seafront are most at risk from the sea, either by erosion or flooding. In seaside towns like Herne Bay the most valuable properties are often found along the seafront. Some of these will be residential properties, but there are also likely to be commercial properties such as hotels, shops and restaurants.

The easiest way to count the number of properties is to walk along the seafront and mark them on a map. Another method is to do a virtual count of the properties on the seafront, using Google Street View. Figure K shows the seafront in Herne Bay and some of the properties along Central Parade.

You can find out the average value of properties on any street in a town from property websites like Zoopla. (This is secondary data, rather than primary data that you gather yourself.) The average price of properties on Central Parade, along the seafront in Herne Bay, was about £650,000 in early 2020. You can check what it is now.

K Properties on the seafront in Herne Bay

Activity

7 a) Count the number of properties along Central Parade in Herne Bay. You can do this using the section you can see on the map in Figure F on page 49, or you could use Google Street View. Based on the average value of £650,000 in early 2020 (or the average now if you have researched it), calculate the total value of the properties on Central Parade (multiply the number of properties by the average value).

b) If you did your own fieldwork, draw a map to show the types of coastal protection for your seaside town. Use Figure L on page 54 as an example to guide you. Calculate the total cost of coastal protection.

c) When a property floods, it is the ground floor or basement that is most likely to be damaged. The average height of the properties on Central Parade is three floors, which means that the cost of flood damage would be about one-third of the total value. Calculate the possible cost of flood damage (divide the total value of properties by the number of floors (in this case, 3)).

Mapping types of coastal protection

In fieldwork techniques 1 and 3 you drew a map, similar to the one in Figure L, to show types of coastal protection and the distribution of property. A map helps you to measure each type of coastal protection using the scale and to count the seafront properties. You will also use a line graph and a bar chart to present your data on longshore drift.

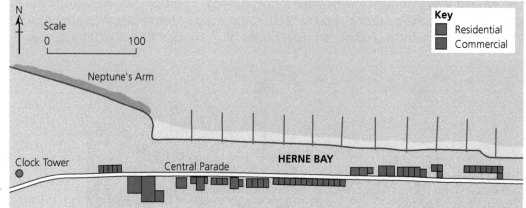

L Coastal protection and seafront properties in Herne Bay

Line graphs

A **line graph** is often used to show a trend over time. The horizontal x-axis is used to show time and the vertical y-axis is used to show another variable, such as distance. The graph is plotted as a series of points, which are then joined with straight lines to create a continuous line showing the trend. The ends of the graph do not have to reach the axes.

Step 1

Draw a graph with two axes – the horizontal x-axis showing time in minutes and the vertical y-axis showing distance in metres. You can use Figure M to help you draw your graph. In this case, because the orange can move to the west or east, the 0 (zero) point on the x-axis is drawn in the middle, with minutes increasing in either direction – west and east of the 0 point.

Step 2

Plot points on the graph to show how long it takes the orange to move from the start point at 0, using the y-axis to help you to show the distance it moves along the beach. It may also help to divide the minutes on the x-axis into 10-second intervals so that you can plot the points accurately.

Step 3

Join the points you have plotted with straight lines to create a continuous line, showing the movement of the orange along the beach. Your line may go west or east from the 0 point, depending on which way the wind blew the orange.

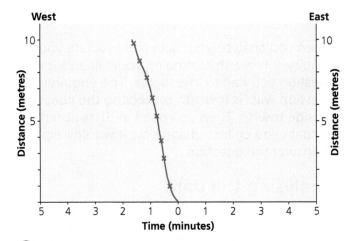

M Line graph to show the movement of the orange along a beach

10 Figure M shows the movement of an orange along a beach, based on the data in the table in activity 5 on page 51.
  a) Using the data in the line graph, which direction was the orange moving – east or west?
  b) How long did it take orange to move 10 metres?

11 If you did your own fieldwork, draw your own line graph to show the movement of the orange along the beach that you measured. If you used two oranges, you can either draw two lines or create an average.

Bar charts

The bar charts in Figure N show the level of the beach on either side of a groyne at five different sites along the beach.

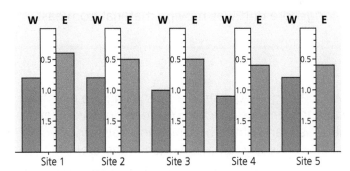

N Level of the beach on either side of a groyne, from the top down

Activities

12 Study the bar charts in Figure N and use the data to complete this table. The first row has been done for you.

Site on groyne	Level of beach on the west from the top of the groyne	Level of the beach on the east from the top of the groyne
Site 1: close to sea wall	1.2 metres	0.4 metres
Site 2: 5 metres from sea wall		
Site 3: 10 metres from sea wall		
Site 4: 15 metres from sea wall		
Site 5: 20 metres from sea wall		

13 If you did your own fieldwork, draw bar charts to show the beach level on either side of the groyne that you measured.

Telling a geographical story: analysing and evaluating your enquiry

◯ **Herne Bay in the summer**

When you analyse your data and evaluate your enquiry, it is worth reminding yourself about the question you had to investigate. The enquiry question was **'Is it worth protecting the coast in a seaside town?'**. Then you need to think about all the data you collected and how it would help you to answer the question.

Analysing the data

First, you identified and measured the types of coastal protection in the town that you visited. You then used the data to calculate the total cost of protection. In Herne Bay, along the section we measured, we found a sea wall, rock armour and groynes. The total cost of protection was just over £6 million, most of which was the cost of the sea wall.

Then, you measured the movement of material along the beach. In Herne Bay we found that longshore drift was moving material from east

Activity

14 a) Plan how you would write an analysis to answer the enquiry question, either using the example of Herne Bay, or using the data you collected in another seaside town. You can draw yourself a large planning grid like the one below and fill in as many of the boxes as you can. Once you have filled in the grid, write up your analysis using the ideas you have written in the table.

Fieldwork technique we used	What was the data we collected?	How does the data help us to answer the question?
Identify and measure different types of coastal protection to work out their cost	In Herne Bay we found examples of a sea wall, rock armour and groynes. We marked them on a map and measured them using the scale.	From the data, we worked out the total cost of coastal protection in Herne Bay.
Measure the movement of material along the beach on the day		
Measure the movement of material along the beach over a long period of time		
Map and count the number of seafront properties and estimate the cost of flood damage		

b) Finally, write a conclusion to answer the question. Don't just answer 'Yes' or 'No'! You should aim to write at least a paragraph and justify your answer with evidence from the data you collected. You could start off by writing something like: *'After carrying out our fieldwork investigation in Herne Bay, I think it is worth protecting the coast in a seaside town. I have come to this conclusion because....'*

to west and that the beach was building up on one side of the groynes. This showed that the groynes were helping to prevent the beach from disappearing.

Finally, you counted the number of seafront properties and, from their average value, you calculated the cost of repairing the damage caused by flooding. In Herne Bay, along the section we counted, there were 50 properties with a total value of £325 million, so the cost of flood damage would be over £100 million.

Evaluating your enquiry

There are two ways to evaluate this enquiry.

If you visited your own seaside town and collected your own data, you can think back on what you did.

If you didn't visit the seaside but instead used the example of Herne Bay, you could evaluate the fieldwork techniques in this chapter and think about what you would do.

P **Students doing fieldwork on a beach**

Activities

15 a) Think back on the fieldwork techniques you used, or the ones you read about in this chapter.
  - How useful were they in helping you to answer the enquiry question?
  - What problems did you/might you experience?
  - What might you do differently next time/if you did it for real?

b) Draw yourself a table like this to write your ideas in. Try to write at least one idea in each box.

Fieldwork technique	How useful was it in helping us to answer the enquiry question?	What problems did we/ might we experience in using this technique?	What might we do differently next time/if we did it for real?
Identify and measure different types of coastal protection			
Measure the movement of material along the beach on the day			
Measure the movement of material along the beach over a long time			
Map and count the number of seafront properties and estimate the cost of flood damage			

16 a) Think of another geographical question that you could investigate at the coast. (It doesn't have to be in a seaside town.)

b) Think of at least two fieldwork techniques that would help you to investigate your question. (You can make up your own techniques, but remember they have to be safe!)

Learning objectives

▶ To understand the factors that cause flooding.

▶ To assess places that are at risk of flooding and choose a fieldwork location.

▶ To analyse flood maps and land use in my local area.

▶ To create an annotated flood risk map.

How do floods happen?

In February 2020, the UK experienced three storms over the course of three weeks. These storms, named Ciara, Dennis and Jorge, gave the UK 141 per cent of its average February rainfall, causing record-high river levels. Heavy, prolonged rain like this can cause high levels of infiltration, highly saturated soil and overland flow running across the land, which can in turn cause devastating river floods. Floods can severely damage our environment, including homes, businesses and farmlands, and can even damage power sources. Many homes and businesses, particularly those built on flood plains, are at high risk. Therefore, it's important that geographers understand the land around them to plan for and protect vulnerable places from flooding. Just as in your coast enquiry in Chapter 5, landscape management is a crucial part of keeping our local areas safe from river flooding.

Activity

In order to understand flooding, you'll need to familiarise yourself with the water cycle. The diagram below shows the water cycle with the arrows showing its different flows.

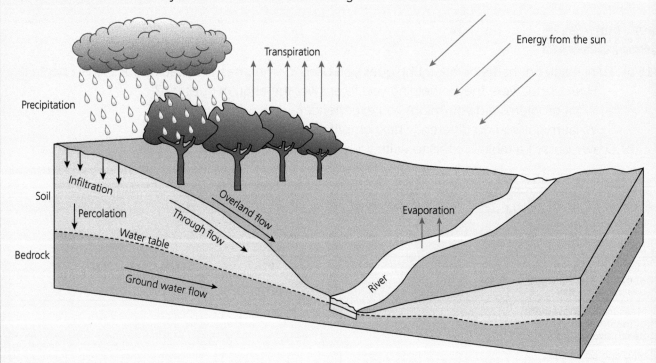

A The water cycle

1 a) Identify which flows (shown by arrows) in Figure A might cause the river to increase in size.

b) Why do the processes of transpiration and evaporation mean that there might be less water in the catchment and river?

c) How might overland flow and infiltration contribute to flooding?

Flood hydrographs

A **flood hydrograph** is a useful diagram that helps you to understand more about how a river changes after rainfall. They show the **river discharge** that occurs as a result of precipitation from an earlier storm. They are used to plan for flood situations as well as times of drought. Figure B shows the flood responses of two large rivers, 'A' and 'B'.

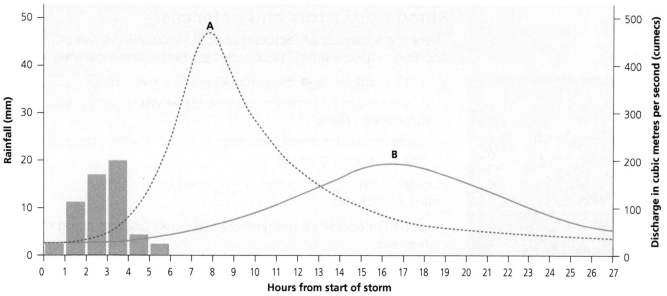

B A flood hydrograph

Activity

2 **a)** Write down the maximum discharge of rivers A and B. This is called the 'peak discharge'.

  b) Work out how many hours there were between the end of the rainfall and the peak discharge for both rivers A and B.

  c) Describe the differences in the shapes of the two response curves, A and B. Which river, A or B, do you think will cause more flooding problems and why?

  d) For the final activity in this section we want you to try and find out why the two rivers (A and B) might have different responses after a rainfall event.

  Complete the table below, adding two or more physical and human factors that might cause these different responses, with explanations. We have provided an example of a physical factor to help get you started.

Human factors	Physical factors	Explanation – why might these have an impact on the river's response?
	River A might be on a flood plain	

C High water levels after a heavy rainfall

Collecting data

Before diving into our fieldwork data collection, it is important to understand the factors that increase an area's vulnerability to flooding, how we can defend against the risk and how we can use fieldwork techniques from your classroom to complete your enquiry.

Flood risk factors and defences

There are a number of factors that could increase the risk of flooding in a local area. These include whether the local area:

- is at a confluence ➔ the place where two rivers meet
- is surrounded by tarmac and concrete ➔ which increases surface run-off into rivers
- has experienced heavy, prolonged rainfall ➔ which saturates the area surrounding a river.

In order to protect vulnerable areas, flood management defences include:

- embankments – permanently raised banks of earth along the river side
- weirs – these control the flow rate of a river during periods of high discharge
- dams – barriers that hold back water or control its flow
- trees – these slow down the rate at which water flows into a river, and can intercept rainfall
- flood walls – walls built along a river's banks to stop flooding.

In the next section you will use this information, together with maps, to examine areas of high risk and select a vulnerable location for your fieldwork.

D Flood defences can be seen on the river in Bewdley

Choosing a suitable fieldwork location

In any geographical enquiry, selecting the right locations is an important first step. You have to make a selection because rivers cover a lot of ground: the River Thames is 215 miles long, for example!

One vital piece of data geographers need before starting a flooding enquiry is a flood map. Flood maps show you where an area might be at high, medium or low risk of flooding. They do this by using a key – white for zone 1 (low risk), light blue for zone 2 (medium risk) and dark blue for zone 3 (high risk). They are a kind of map known as a choropleth map, which means they use different shades of a colour to show different values or numbers. Flood maps also show you where flood defences are already in place, using an orange line. It is important for you to select specific sites to focus on for your fieldwork. Flood maps are created by the Environment Agency: therefore you will be using data someone else has collected – secondary data. You will explore other kinds of secondary data in this chapter and elsewhere in this book.

Step 1

Use the government's Flood Map for Planning website https://flood-map-for-planning.service.gov.uk/ to find out if a local area near your school is at risk of flooding (in flood zone 1, 2 or 3). You will need to search by postcode.

Step 2

Explore the area by zooming out to look for examples of flood defences, the different levels of flood risk and where the river is. Find your school and place the yellow marker onto it as a reference point.

Step 3

Zoom into the map as far as it will go and print the map out (in colour, if you can). Split your map into nine square grids – three going across and three going down, using a pen and a ruler. Label each square with a number (1–9). These grids will be the nine locations you will consider when choosing three suitable fieldwork sites.

Step 4

Examine each square closely. Choose the three grid areas that you think are the most vulnerable to flooding to investigate in more detail. These will be your three fieldwork sites.

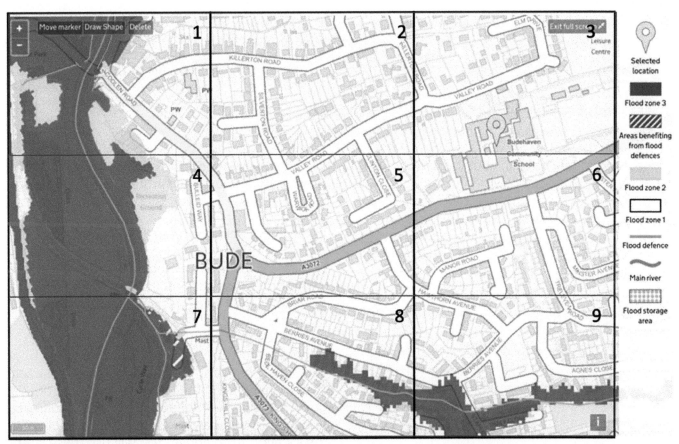

E This flood map from the government's Flood Map for Planning service shows Bude, in Cornwall.

Collecting data

Fieldwork technique 1: Flood risk assessment

A **flood risk assessment** is a fieldwork technique that allows you to find out how vulnerable an area is to flooding. You may have heard the term 'risk assessment' before in fieldwork – it is an important way to assess the possible hazards present in fieldwork enquiries to keep you or the area around you safe. In order to complete a flood risk assessment, you can either go out to visit your chosen site from the ones selected in the location assessment, or you can complete it using online maps and databases.

One limitation of the location assessment is that it doesn't provide you with information on the current level of the river. You will need this data as part of your flood risk assessment. In order to get this secondary data, visit https://riverlevels.uk/map

and put the name of your local area or river in the search bar to the top left of the screen. You'll be given a selection of monitoring stations nearby, so you'll need to choose the station closest to your fieldwork locations.

The hydrographs at the bottom will show the measurement of the river over the previous week, and over a longer period of time. The yellow line is the typical low reading, the green is the typical high reading, and the blue is the current level of the river. If the blue line is above the typical high, flooding is possible. The hydrograph also provides data on previous flooding events. To view this, drag the slider along the long term hydrograph to investigate if there have been any past floods.

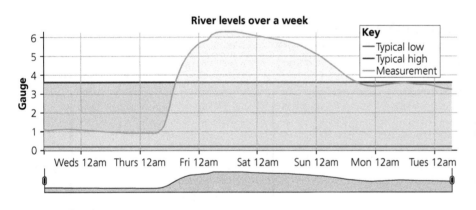

F Measurement of river levels of the River Don at Doncaster over a week in November 2019

Step 1

Collect secondary data on your local river level over the past year by visiting the https://riverlevels.uk/map website. Before analysing it, print a copy – you will use this later in your data presentation. Looking carefully at the data, complete the following table to add vital information:

River name:

Highest river level (what was it and when?)	
Lowest river level (what was it and when?)	
How many times a month was the river over its typical high?	
Which month had the highest river levels?	

Step 2

Create your own flood risk assessment fieldwork sheet using the example in step 1 as a template.

Step 3

Locate your fieldwork area. If you are completing this fieldwork in the classroom, use the satellite and terrain features on Google Maps to complete the flood risk assessment (these can be selected by clicking the three lines next to the search bar on Google Maps). If you are able to go out and complete the flood risk assessment in person, remember to take photos of the evidence you see of flood risk and flood defence. Explore each site carefully, using the flood risk assessment as a guide to help you find factors that might increase flood risk. If you can, take some screenshots of the location from above. You will use these later in the chapter.

Step 4

Using your in-person exploration of your fieldwork site, or your virtual version on Google Maps, score each risk factor from 1 to 5 on your flood risk assessment sheet for each of your three sites. A low score on the scale of flood risk indicates a low risk of flooding. For example, if the river isn't higher than it is normally from your hydrograph, or if there are visible flood defences in place, you might give it a low score of 1 or 2. In comparison, a high score would indicate a high risk of flooding. For example, if there are no green spaces and lots of housing, car parks or concrete which would increase surface run-off into rivers after rainfall. Count the total score and put it at the bottom of each sheet.

Flood risk assessment

Date:

Location and site number:

Flood risk score
1 = very low, 5 = very high

Low flood risk factor	1	2	3	4	5	High flood risk factor
Risk of flooding on flood map is zone 1						Risk of flooding on flood map is zone 3
Flood defences in place						No flood defences in place
Low risk of surface run-off (e.g. lots of soft, porous surfaces such as large green spaces)						High risk of surface run-off (e.g. lots of hard surfaces such as concrete or roads)
Residential or town centre far from river						Residential or town centre close to river
Far away from a steep valley						Close to a steep valley
No confluence of rivers nearby						Close to a confluence of rivers
Total risk level				/30		

Activity

3 **a)** Give two reasons to explain your choice of fieldwork sites from the flood map (either the one on page 61 or the flood map for your own area). Use data from the map in your explanation.

   b) Look at your local river hydrograph. What was the highest river measurement in the past year?

   c) List the kinds of flood defences you observed during your fieldwork. If you did not see any, which flood defence might be suitable for your fieldwork sites and why?

   d) What were the main differences between the information provided in the flood map and the information you could get from a Google Maps satellite or terrain view? List at least three differences.

Flood risk bar chart

Another creative way of presenting your fieldwork data is to use colours and apply them to charts.

Using your flood risk assessment sheet, colour in the scale section. You could use a deep red to show high risk, orange for medium risk, and green for low risk. This will turn your assessment sheet into a horizontal bar chart.

Flood risk assessment

Date: 10/2/20

Location and site number: Site 1 (Bude, River Neet, between Bencoolen Road and Vicarage Road)

Low flood risk factor	1	2	3	4	5	High flood risk factor
	Flood risk score 1 = very low, 5 = very high					
Risk of flooding on flood map is zone 1						Risk of flooding on flood map is zone 3
Flood defences in place						No flood defences in place
Low risk of surface run-off (e.g. lots of soft surfaces such as large green spaces)						High risk of surface run-off (e.g. lots of hard, surfaces such as concrete or roads)
Residential or town centre far from river						Residential or town centre close to river
Far away from a steep valley						Close to a steep valley
No confluence of rivers nearby						Close to a confluence of rivers
Total risk level			/30			

Annotated flood risk map

One way of presenting your own data of flood risk is to bring together all the information into a detailed annotated map. Annotated maps can include primary and secondary data, as well as showing fieldwork locations if you visited multiple sites.

Step 1

Using a large A3 sheet of paper (or bigger), stick the flood risk map you printed earlier into the centre of the paper. Identify on the map the three sites you used in your fieldwork enquiry by drawing or marking them on the map and clearly labelling them.

Step 2

Now it is time to build your annotated map with all the data you have collected. Stick the printed copy of your flood hydrograph onto the bottom of the map and label it with the river name and the time period the data was collected.

Step 3

Starting with site 1, find a good space on the outside of the map to stick one or two photographs taken during fieldwork – or screenshots from the internet – of the site. Choose photos that you think best show flood risk or flood defence. Avoid adding too many photographs at this stage, as you will need space to add more pieces of data and annotations.

Step 4

Annotate the site and the photograph with key flood risk information. These may be examples of residential or commercial areas, riverside properties or green spaces that you can show in the photographs or indicate on your flood map.

Step 5

Cut out and stick on your flood risk assessment bar chart for site 1.

Step 6

Follow steps 1–5 again for the other fieldwork sites to complete the annotated map.

Date: 8 March 2020
Location and site number: Site 3 (Hereford Old Bridge)

Low flood risk factor	1	2	3	4	5	High flood risk factor
Risk of flooding on flood map is zone 1						Risk of flooding on flood map is zone 3
Flood defences in place						No flood defences in place
Low risk of surface run-off (e.g. lots of soft surfaces such as large green spaces)						High risk of surface run-off (e.g. lots of hard surfaces such as concrete or roads)
Residential or town centre far from river						Residential or town centre close to river
Far away from a steep valley						Close to a steep valley
No confluence of rivers nearby						Close to a confluence of rivers
Total risk level				19/30		

Flood risk score (1 = very low, 5 = very high)

Date: 8 March 2020
Location and site number: Site 1 (A438 Lugg Flats)

Low flood risk factor	1	2	3	4	5	High flood risk factor
Risk of flooding on flood map is zone 1						Risk of flooding on flood map is zone 3
Flood defences in place						No flood defences in place
Low risk of surface run-off (e.g. lots of soft surfaces such as large green spaces)						High risk of surface run-off (e.g. lots of hard surfaces such as concrete or roads)
Residential or town centre far from river						Residential or town centre close to river
Far away from a steep valley						Close to a steep valley
No confluence of rivers nearby						Close to a confluence of rivers
Total risk level				14/30		

Flood risk score (1 = very low, 5 = very high)

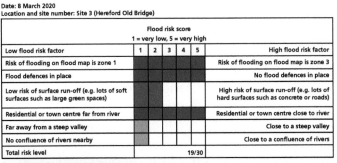

Site 3
- Historic city centre
- Riverside properties
- Flood defences

Site 1
Road liable to flooding 200 yds
- Farmland / nature reserve
- Floodplain for River Lugg
- A438 raised up flood defence

Site 2
- Residential / commercial
- Located along Eign Brook (stream)
- No flood defences

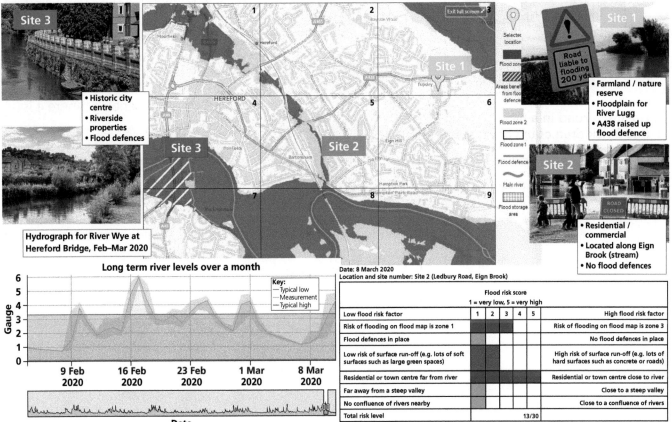

Hydrograph for River Wye at Hereford Bridge, Feb–Mar 2020

Long term river levels over a month

Key:
— Typical low
— Measurement
— Typical high

Gauge (y-axis: 0–6)
Date (x-axis): 9 Feb 2020, 16 Feb 2020, 23 Feb 2020, 1 Mar 2020, 8 Mar 2020

Date: 8 March 2020
Location and site number: Site 2 (Ledbury Road, Eign Brook)

Low flood risk factor	1	2	3	4	5	High flood risk factor
Risk of flooding on flood map is zone 1						Risk of flooding on flood map is zone 3
Flood defences in place						No flood defences in place
Low risk of surface run-off (e.g. lots of soft surfaces such as large green spaces)						High risk of surface run-off (e.g. lots of hard surfaces such as concrete or roads)
Residential or town centre far from river						Residential or town centre close to river
Far away from a steep valley						Close to a steep valley
No confluence of rivers nearby						Close to a confluence of rivers
Total risk level				13/30		

Flood risk score (1 = very low, 5 = very high)

G An example of an annotated flood risk map

We already know that good analysis is all about making sense of the geography that you have uncovered. The next step in your enquiry is to analyse the data you have collected and presented.

Analysing flood risk

The map and flood risk bar charts you created help to visualise the areas at most risk. Using the additional data you have collected on river levels and the land use around your chosen sites, think about the following questions.

H Having a road in a high flood risk area could have economic impacts

Activity

4 a) Describe two ways the physical landscape around each site could impact on the site's flood risk.

b) Which individual risk factor scored the highest across all three sites?

c) Consider one reason why this risk factor scored highest at each site.

d) Using your annotated map, river level data and flood risk assessment, which of the three areas you selected do you think is most at risk of flooding? Explain your answer giving:
- the site you believe is most at risk, describing where it is using locations and compass points
- at least three examples of why
- at least three examples of data to support your answer
- a comparison between two or more sites.

e) Create a table like the one below to show some of the effects flooding will have on the three sites you observed during your fieldwork. Use the three categories given in the table below – one example is done for you.

Site 1: Bencoolen Road, a main road

Economic	Social	Environmental
Bencoolen Road, a main road, is close to a high-risk flood area	People's homes on Swallow Close are on a flood plain – homes and belongings could be destroyed	Ecosystems are flooded causing stress to both plants and animals – the recreation ground is in a zone 2 flood risk area

Evaluating flood risk fieldwork

Doing fieldwork, both in person and online, comes with its challenges. In the next section, you will use your reflection skills to answer questions on the limitations of choosing a fieldwork location, completing physical geography fieldwork from a distance and what else you could have added to your enquiry.

Activity

5 a) What were some of the challenges you found when exploring flood risk areas using Google Maps?

b) Reflect on the sites that you chose for your fieldwork, splitting the flood map into sections and choosing the three most at risk areas. Looking back, do you think this was the best approach? Give at least one example of a limitation of this approach.

c) If you were to go to a river to conduct fieldwork, select the health and safety items or research you might need from the following options:

Maps	Plimsols	Mobile phone	Rope	Waterproof clothing
Safety pins	High vis jacket	Weather forecast	Sturdy outdoor shoes	Sunglasses

Explain how each item is helpful for health and safety during flood risk fieldwork.

d) Thinking back through what you have learned in this book and previous fieldwork you have conducted, give one other fieldwork technique that you could use for this enquiry and explain how you would use it.

e) Explain how each of these maps could help you in your enquiry:
- a topographic map showing contours
- a GIS (Geographic Information System) map showing previous floods in the area
- an economic map showing house prices in the area
- a geologic map showing the type of rocks on the land
- a weather map showing predicted types of weather.

f) Were the flood assessments accurate in identifying areas that flood, and if not, how could they be improved?

Properties along the River Wye could be at risk of flooding

Learning objectives

▶ To understand what factors affect people's quality of life.

▶ To understand how primary and secondary data can be used to measure quality of life.

▶ To design my own quality of life questionnaire.

▶ To analyse and evaluate results from graphs, radar graphs, compound bar charts and coded responses.

What is quality of life?

Quality of life means the general wellbeing of individuals and societies. In geography, it is often linked to development. Development can be measured in different ways, including the Human Development Index, or HDI. HDI, calculated by the United Nations (UN), measures average life expectancy, level of education and income for each country. According to the UN, Norway topped the HDI quality of life global ranking in 2019, with the UK ranking 15th in the world.

A An aerial view of Bergen, Norway

Like development in countries, quality of life in places can be measured in a number of ways. Every year, councils around the UK ask their residents to complete a 'quality of living' survey. Below are some of the key indicators that are used to measure quality of life in a particular area:

- Housing – is the housing made to a good quality? Is there affordable housing?
- Transport – is the area well served by public transport or is it difficult to reach?
- Shopping – do people have access to high quality shops and services, e.g. a shopping mall or a healthcare centre?
- Environment – is the quality of the environment good or bad?
- Leisure – are there lots of activities to do, and facilities for them, such as leisure centres?
- Safety/crime – do people feel safe where they live? Are there any CCTV cameras?
- Community – do people have a strong sense of community and feel connected to place?
- Employment – are there lots of places to work?

Activity

1 a) Which indicator of quality of life do you think the most important and the least important out of the list?

People's quality of life can vary depending on where they live.
The maps below show two very different places.

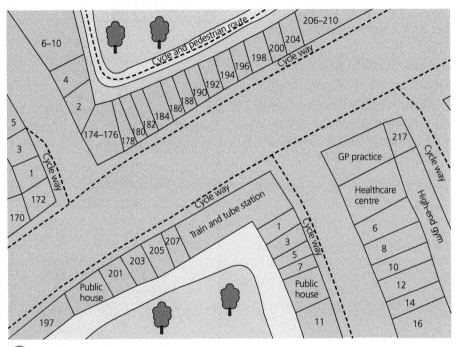

B Map A

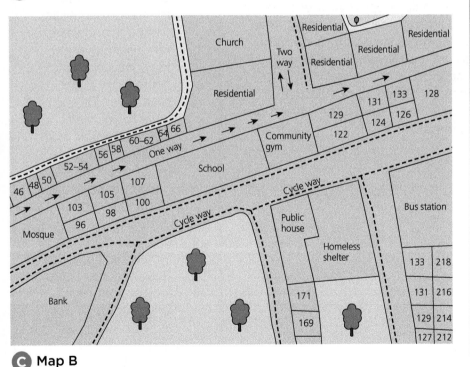

C Map B

Activity

1 **b)** Get into pairs. One of you pick Map A and the other pick Map B. Using just the information given to you on the map, discuss with your partner: which one gives people the better quality of life, and why? Give a list of examples from your map and use as many of the factors of quality of life as you can.

 c) Give an example of two areas local to your house or your school. They should be very different to one another, but close to one another.

In order to answer the enquiry question, you can use both primary and secondary data to find out more about some of the factors that affect quality of life: employment rates and environmental quality.

Fieldwork technique 1: Census statistics

The previous enquiry on flooding introduced you to flood maps, which are a type of secondary data – data collected by someone else. Another source of secondary data that geographers use is statistics. You may have seen the use of statistics in the media, on news reports for example. Many of these secondary data sources come from the Census.

11.45 12 Jan

Unemployment in Wales rises above UK average

Jobless rates rise by 4.1% with a rise of 30,000 jobless.

Read more >

D **Statistics in the media often come from the Census**

Secondary data, such as statistics, can help support your enquiry by looking at some of the quality of life indicators in more detail. One indicator that you could use secondary data for is unemployment rates in your region. If there is high unemployment in an area, the quality of life may be lower because those who are unemployed may not be able to access other aspects of quality of life, such as affordable quality housing, due to cost. Plus, being unemployed can affect people's wellbeing and happiness.

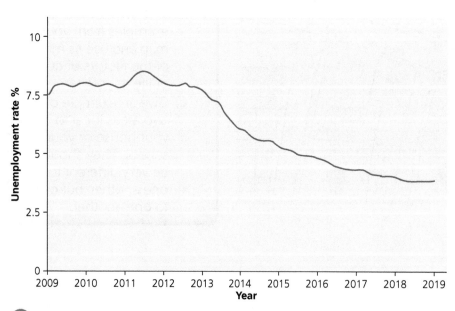

E **The unemployment rate measures the numbers of people aged 16 and over who are not in work**

Activity

2 Figure E shows the unemployment rate in the UK since 2009.
   a) Which year was unemployment at its lowest?
   b) Describe when unemployment was at its highest.
   c) Describe the trend of the line on the graph.
   d) Why do you think high unemployment might mean a lower quality of life?
   e) Collect unemployment data over the past ten years for your area. Go onto the Office for National Statistics webpage and search for unemployment in your local area or nearest city in the search bar. On the table to your left tick 'time series' and 'dataset', and you will be given a range of options to select from. Choose an option that gives 'all' – which will give you male and female data together. Now, you can either look at the graph, or select 'show data as table' which will give you the exact unemployment figures for each year.

Fieldwork technique 2: environmental quality survey

In addition to secondary data, this enquiry will involve you collecting primary data on the quality of the environment. The characteristics of the environment around you are another important factor of quality of life. This includes the houses people live in, the litter you might see on the ground, and how safe you feel. Generally, the higher the quality of housing, the cleaner the streets, and the safer people feel, the greater the quality of life. Geographers studying the environmental quality of an area often use a survey known as an environmental quality survey. You can use an environmental quality survey to go out and assess the environmental quality of your local area.

Environmental quality survey						
Location: Time:						
	Good +2	OK +1	Av 0	Poor −1	Bad −2	
Vibrant, interesting place						
No vandalism						
						No shops
						Dirty and unhealthy, e.g. smelly, litter
No traffic noise						
						Dangerous for cyclists, e.g. no cycle lanes

F Environmental quality survey template

Each aspect of environmental quality is put on a special bipolar (or opposites) scale ranging between +2 and −2. You may have seen this bipolar scale before in enquiry 2, when creating your sustainable community assessment.

As you can see, the left-hand side of the survey is the positive end of the scale and shows examples of good environmental quality. For example, if the general area is vibrant and interesting, you may tick the +2 (good) or +1 (OK) box, depending on your personal opinion. On the right-hand side are negative scores. Have a look around your area and judge whether you think it should have a positive or negative score. At the end, count everything up and give it an overall score.

Activity

2 **f)** Some of the examples on the bipolar survey above are missing. Using the opposite side of the scale as a guide, fill in the blanks to best describe each area. For example, the opposite of a 'vibrant, interesting place' might be a 'boring, empty place'.

g) Give one reason why it is important to include the general description of the site you are visiting.

Collecting data

Fieldwork technique 3: quality of life questionnaire

Questionnaires are a great way of gathering valuable primary data in fieldwork. One way to investigate quality of life in your local area is to use the eight factors listed on page 68 and ask respondents to rate each one. It's up to you which scale you wish to use. The example below uses a scale of 1–5, where 1 is poor and 5 is good. This kind of range is familiar – many people may have completed satisfaction surveys when they've been to an event or have rated a product they've bought online.

Location:					
Time:					
On a scale of 1–5, where 1 is poor and 5 is good, how do you rate these aspects of quality of life in this area?	1	2	3	4	5
Housing					
Transport					
Shopping					
Green space					
Leisure					
Safety/Crime					
Community					
Employment					

G Quality of life questionnaire

This response, where people score the factors influencing quality of life on a 1–5 scale, is an example of using closed questions. Another example of closed questioning might be a question which just has a 'yes' or 'no' answer.

It is also a good idea to include more open questions, where respondents can answer in more words or explanation. An example of an open question might be 'What is the best thing about living in this area?' Open questions provide interesting and unique answers based on people's personal opinion. For example, someone may score 'leisure' as 5, but you later find out from another person that they think it's the worst thing about living in the area, because the new leisure facility is too expensive.

Here is an example of some open question boxes for a quality of life questionnaire:

What would you say is the...	
...best thing about living in this area?	
...worst thing about living in this area?	

H A student using a questionnaire to conduct primary research

Creating your quality of life questionnaire

On a sheet of A4 paper, or using Word on a computer, create your quality of life questionnaire.

Step 1

Give your questionnaire a title. Include a line for 'Location', so that you can fill in your precise location when you are out in the field. Add a line for 'Time', so that you can record what time of day you are out doing your fieldwork.

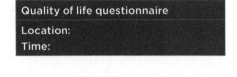

Step 2

Draw a table under your title, location and time headings following the environmental quality survey template on page 71. You should have rows for all the quality of life indicators you are using for your enquiry.

Good	OK	Av	Poor	Bad
+2	+1	0	–1	–2

Step 3

Underneath, draw the open questions boxes shown on the previous page.

Step 4

Decide how you are going to select people for your study. This is called sampling. You could either:

- choose people randomly (random sampling)
- choose every third person that walks by (systematic sampling)
- choose people from every type of background and from all ages (stratified sampling).

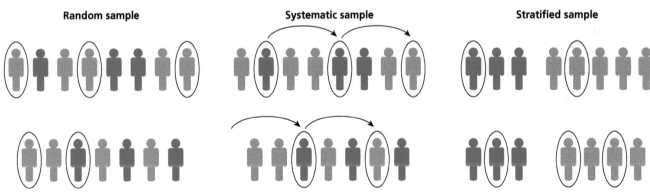

Random sample   **Systematic sample**   **Stratified sample**

🛈 **Three types of sampling techniques**

Activity

3 a) Using the unemployment data you collected in activity 2e), or the example for the North East of England below, create a line graph that shows the changes to unemployment levels over the last ten years.

2009	2010	2011	2012	2013	2014	2015	2016	2017	2018
9.2	9.5	10.8	10.2	10.0	9.1	8.1	7.1	5.6	4.9

J **Unemployment rates in North East of the UK (%) since 2009**

In this section, you will create three types of graphs to present the data you have collected.

Line graphs

A line graph is the best way of showing different trends or changes in data over time. Go back to page 54 to remind yourself about how to draw a line graph.

Radar graphs

A radar graph is a good way to demonstrate data from a bipolar survey. Look back to page 19 to remind yourself about radar graphs.

Activity

3 b) Using the environmental quality survey data provided below, or data from your own survey, create a radar graph. Use the radar graph outline below to help you get started.

Environmental quality survey						
Location: Salford Shopping Centre **Time: 11 a.m.**						
	Good +2	OK +1	Av 0	Poor −1	Bad −2	
Vibrant, interesting place	✓					Empty, boring place
No vandalism			✓			Badly vandalised
Wide variety of shops	✓					No shops
Clean and healthy, e.g. fresh air, no litter				✓		Dirty and unhealthy, e.g. smelly, litter
No traffic noise					✓	Loud traffic noise
Safe for cyclists, e.g. cycle lanes				✓		Dangerous for cyclists, e.g. no cycle lanes

K Radar graph outline

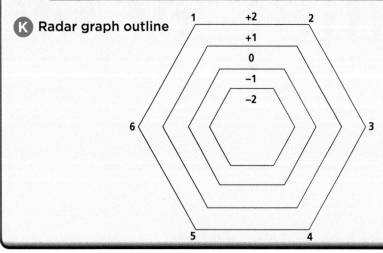

Compound bar charts

A compound bar chart is a good way to present the first part of the questionnaire data that you collected out in the field.

Step 1

Create your x- and y-axes. The x-axis will be where you create your bars for each of the quality of life categories, so be sure to have enough space (1 cm) to create the bar, and enough space in between so that each bar is not touching its neighbour. Label each section with one of the quality of life factors. To create your y-axis, count the number of participants questioned to see how high the axis should go.

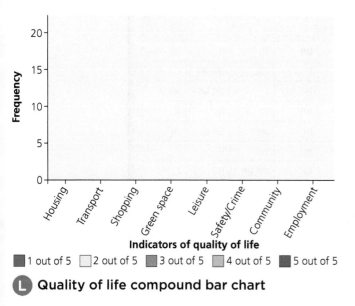

1 out of 5 ☐ 2 out of 5 ■ 3 out of 5 ☐ 4 out of 5 ■ 5 out of 5

 **L** Quality of life compound bar chart

Step 2

Select a colour to represent each number of your scale. For example, in the compound bar chart above, we have chosen red for every score of 1, yellow for 2, green for 3, orange for 4 and blue for 5. Ensure you have a key for your chart to show what these numbers represent.

KEY:
Score out of 5

■ 1 out of 5 ■ 3 out of 5 ■ 5 out of 5
☐ 2 out of 5 ■ 4 out of 5

M Key showing score out of 5

Step 3

Now it's time to start adding your data to your bar chart. Starting with your first category (in this example, housing), add how many people gave housing a score of 1 and colour it in with the correct colour from your key.

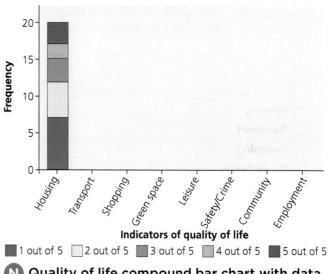

■ 1 out of 5 ☐ 2 out of 5 ■ 3 out of 5 ☐ 4 out of 5 ■ 5 out of 5

N Quality of life compound bar chart with data

Step 4

Count how many people gave your first category a '2'. Stack this number on top of your bar. For instance, if two people gave housing a 1, and two people gave it a 3, stack three on the top of your previous bar. Continue this process until you have filled your first bar. Now move on to your second factor and continue the same process of stacking the data on top of one another.

During the quality of life questionnaire, you asked people to describe what they think the best and worst thing about living in this area is. Your first job is to have a look through this data and **code** the responses. Coding helps us find common themes in the data. For example, coding people's responses into a table can show us that most people find shopping the best thing about living in your area, and crime the worst. In the next exercise, we are going to code the responses people gave into our eight quality of life themes.

Activity

4 Using your own data (or the data provided for Salford below), create your own compound bar chart to present quality of life data.

On a scale of 1–5, where 1 is poor and 5 is good, how do you rate these aspects of quality of life in this area?	1	2	3	4	5
Housing	LHT II	LHT	III	II	III
Transport	I	II	IIII	LHT LHT	III
Shopping		II	I	LHT LHT	LHT II
Green space	LHT III	IIII	III	II	II
Leisure	III	III	LHT	LHT II	III
Safety/Crime	LHT LHT	IIII	IIII	II	
Community	I	IIII	LHT LHT	IIII	I
Employment	LHT II	LHT II	III	III	

◎ **Quality of life data for Salford Shopping Centre**

Analysing your quality of life questionnaire

Step 1

Draw a table with all the different factors that influence quality of life:

Best things about living in Salford

Housing	Transport	Shopping	Green space	Leisure	Safety/Crime	Community	Employment

Worst things about living in Salford

Housing	Transport	Shopping	Green space	Leisure	Safety/Crime	Community	Employment

Step 2

Look through the responses from your questionnaire. Place each answer into the box that you think best describes it, in both the corresponding 'best' and 'worst' tables. For example, for the 'best thing about living in this area', it might look like:

Best things about living in Salford

Housing	Transport	Shopping	Green space	Leisure	Safety/crime	Community	Employment
Modern apartments Cheap homes	Fast trains Tube	Westfield	Nice park	Big new gym	Neighbourhood watch scheme	Friendly neighbours	Good pay

Worst things about living in Salford

Housing	Transport	Shopping	Green space	Leisure	Safety/crime	Community	Employment
Too expensive	Overcrowded trains	Small supermarkets	Often full of dog mess and litter	New gym is £165 per month	Lots of muggings Pubs are scary at night	I don't know anyone in my block	Few job centres Few full-time jobs – all part time

Analysing all of your data together

Look at all the data you have presented, either through your fieldwork for your own local area, or for the data provided in this enquiry, and answer the following questions:

Activities

5 a) What were the highest scoring aspects of environmental quality in your radar graph?
   b) What were the lowest scoring aspects of environmental quality in your radar graph?
   c) Calculate the mode of your environmental quality survey (the number that appears most frequently in data).
   d) What does the mode tell you about the overall environmental quality?
6 a) Look at your compound bar chart. Describe the two highest scoring aspects of quality of life.
   b) Describe the two lowest scoring aspects of quality of life.
   c) Choose two other factors of quality of life and compare their results to each other. How did people score them? How many people gave them 5 out of 5 and how many gave them 1 out of 5?

7 a) Analyse your qualitative results from coding. What are some of the common themes? What do people tend to say frequently about the best and worst aspect of living in your local area (or in Salford)?
   b) In your opinion, what is the worst thing that someone said to describe living in your local area (or in Salford)?
   c) Compare this with the scores for that category in your compound bar chart. Does the bar chart agree?
8 Compare the environmental quality survey data to the responses you coded into the category 'environment' and your compound bar chart on 'environment'. Is there is a relationship between your environmental quality survey and what people said in your questionnaire data?

Evaluating your unemployment data

In this investigation, you have collected a mix of qualitative and quantitative data, primary and secondary data, used line graphs, radar graphs, compound bar charts and coded responses to questionnaires. You need to evaluate if there could be any issues with your data that could affect how you have analysed it.

Activity

9 What issues can you see with the two graphs, Figures P and Q?

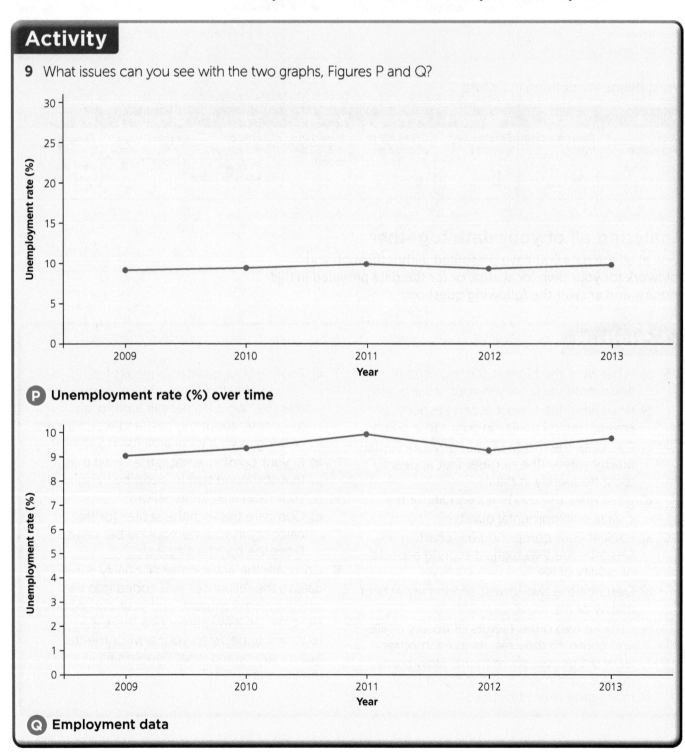

P Unemployment rate (%) over time

Q Employment data

Evaluating your environmental quality survey

If you compare your results to your neighbour, you may have very different results. This is because any survey you complete on your own is subjective – meaning that it is based on your own personal opinion. This means that results can vary significantly. Even our opinion on what makes an area 'vibrant and interesting' looks very different.

Evaluating your questionnaire

It is often difficult to complete questionnaires. If you live in a busy city, you may feel that no one will stop to speak to you, no matter how much you ask! This is all part of the process of fieldwork – it doesn't always happen the way you planned.

Earlier in the unit (page 73), you were asked to choose a way of selecting people for your questionnaires. This is known as the way you 'sample' your respondents – whether that be randomly, systematically or stratified. Thinking back to your own study, do any of the following speech bubbles describe your experience?

Systematic

> I tried to pick every 'third' person that walked past me, but they all ignored me.

Stratified

> I wanted to approach people from different backgrounds and ages, but in truth, everyone who lives in my area is really similar.

Random

> Even when I chose people randomly, I still felt that I was biased towards certain people, like older people, who were more friendly.

Activity

10 a) Think of another limitation of your environmental quality survey which may have influenced your results.

b) How did your sample – the way you selected people to ask questionnaires – affect your survey?

c) Consider the speech bubbles on the left or your own reflections on a questionnaire you may have done out in your local area. Give two reflections on what you found difficult during your questionnaire, and how this may have affected your results.

d) Consider how the number of people (sample size) you use might affect your results. Is it better to have a large sample size, or a small one? Explain why.

e) Instead of completing a questionnaire, you may wish to use an online survey to send out. Give one advantage and one disadvantage of using online surveys.

Learning objectives

▶ To analyse population data.
▶ To conduct a resident questionnaire.
▶ To design a population pyramid.
▶ To interpret the population of my local area.
▶ To evaluate population data.

What is population?

The number of people living in a particular place is known as the population. In this enquiry, you will collect data on how the local population of your area has changed over time.

In the previous enquiry, you learnt about the Office for National Statistics (ONS) and the Census (page 70). To prepare for this enquiry, Census data will be useful, as it provides detailed information on the population of any given place or area.

DataShine is a resource that geographers use to explore some of the most recent Census data in different places. It includes categories like health, housing and education. Geographers use maps like these to help visualise how areas differ. In fact, you have already seen a chloropleth map like the one below in Chapter 6. DataShine can be accessed for free online at www.datashine.org.uk.

4.9% of people have lived in the UK for 10 or more years

26.7% of people have lived in the UK for 10 or more years

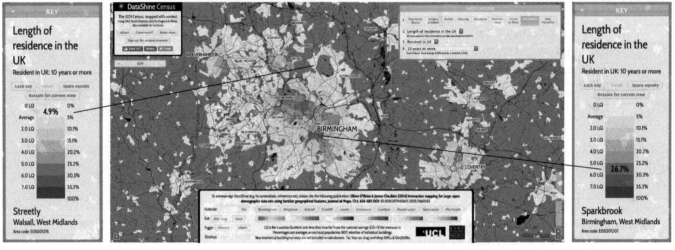

A A DataShine map of Birmingham showing how long people have been resident in the UK

DataShine can help you find out more about how the population in your area has changed. In the next exercise you will explore choropleth maps showing the percentage of people arriving to your local area from outside of the UK.

Step 1

Using a computer, go to the DataShine website and type in your school postcode at the bottom left of the screen.

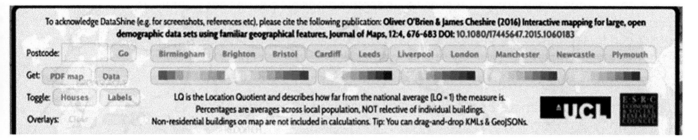

Step 2

At the top of the screen you will see a bar called 'data chooser'. Select 'Residency'. You will be presented with some drop-down options to choose from. Select 'Year of arrival in UK' from the options.

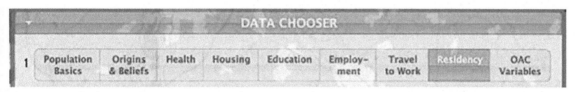

Step 3

You will now be presented with different years. This data shows you what percentage of people arrived in your area in particular years, like between 1941 and 1950 and so on. For this activity, select '1961–1970'. The colours on the choropleth map represent the percentage of people who arrived in that area between 1961 and 1970.

Step 4

Compare that map with a more recent one, for example 'Arrived in 2010-2011'.

Activity

1. **a)** Find and write down three areas near where you live where there is a higher percentage of people who arrived in the UK from 1961–1970.
   b) Find and write down three areas near where you live where there is a higher percentage of people who arrived in the UK from 2010–2011.
   c) Using the data (the percentages of people who came into the UK in those years) describe how the data for 1961–1970 and 2010–2011 is different.
   d) Select another option on the data chooser and explore the data in your area. Write down an example of something you found that surprised you about your area.

Collecting data

The size of your local population may surprise you. In population fieldwork, using secondary data will give you a better understanding of what your area looks like and how it has changed over time. However, to understand people's experiences of living in your local area, secondary data isn't enough by itself. In this section you will be introduced to secondary data from the ONS and you will also learn how to create a residents' questionnaire to gather your own primary data.

Fieldwork technique 1: Secondary data from the ONS

In this section you will focus on population data taken from the ONS. The ONS collects population data across nine key regions in England and breaks this data down into age bands.

The ONS is best known for their once-in-a-decade Census survey which gives the most accurate estimate of the people and households in England and Wales. However, they also give future projections, or estimates, based on their Census data which is how the data for Figure B has been calculated. The ONS website (www.ons.gov.uk) provides a range of different types of economic and social data that is regularly collected by a team of researchers. You can go onto the website to explore the kinds of data they have available. Your teacher will be able to help you gather population data from the ONS, but to help you get started, below is a table of recent population data for you to use in your enquiry, showing regional population data for men and women in England.

Age group	North East	North West	Yorkshire and the Humber	East Midlands	West Midlands	East	London	South East	South West
MALES									
0–9	11.8%	12.7%	12.6%	12.2%	12.9%	12.9%	14.0%	12.6%	11.6%
10–19	11.6%	11.9%	12.0%	11.8%	12.3%	11.8%	11.4%	12.1%	11.4%
20–29	14.1%	13.7%	14.1%	13.6%	14.1%	12.0%	15.2%	12.4%	12.7%
30–39	12.2%	12.8%	12.6%	12.1%	12.8%	12.8%	18.9%	12.4%	11.8%
40–49	11.8%	12.4%	12.4%	12.6%	12.5%	13.1%	14.2%	13.3%	12.2%
50–59	14.0%	13.6%	13.5%	13.9%	13.1%	13.7%	11.5%	13.9%	13.9%
60–69	11.7%	10.8%	10.8%	11.1%	10.3%	10.7%	7.4%	10.6%	11.8%
70–79	8.5%	8.1%	8.0%	8.6%	8.0%	8.5%	4.7%	8.3%	9.7%
80+	4.2%	3.9%	4.0%	4.1%	4.1%	4.6%	2.7%	4.5%	5.0%
FEMALES									
0–9	10.8%	11.7%	11.7%	11.4%	12.1%	11.9%	13.3%	11.6%	10.6%
10–19	10.5%	11.0%	11.2%	11.1%	11.4%	10.8%	10.8%	11.1%	10.5%
20–29	12.9%	12.8%	13.2%	12.6%	13.1%	11.1%	15.1%	11.4%	11.5%
30–39	12.3%	12.7%	12.6%	12.3%	12.7%	12.9%	17.7%	12.5%	11.6%
40–49	12.0%	12.5%	12.3%	12.6%	12.4%	13.0%	13.7%	13.3%	12.2%
50–59	14.2%	13.7%	13.4%	13.9%	13.2%	13.6%	11.9%	13.7%	14.0%
60–69	12.0%	10.9%	10.9%	11.2%	10.5%	11.1%	8.0%	10.8%	12.2%
70–79	9.2%	8.8%	8.8%	9.1%	8.7%	9.2%	5.5%	9.1%	10.3%
80+	6.1%	5.8%	5.9%	5.9%	5.9%	6.4%	3.9%	6.4%	7.1%

B Regional population data for men and women in England

Actvity

1 e) Using the ONS website, find three themes that the ONS provides data on, not including the ones mentioned above. Try to find three things that link to one of the other chapters in this textbook. For example, unemployment data links in to Chapter 7 on Quality of Life.

 f) Using the data in Figure B, identify the largest population group in your regional area.

 g) Which region has the highest **total** percentage of 70+ year olds, according to the data in Figure B?

Fieldwork technique 2: residents' questionnaire

Although secondary data sources are a helpful first step in understanding populations, in order to get more insight on your local population, you will find it useful to speak to people in your area about how the population has changed. You can do this by using what's called a residents' questionnaire. In this section, you will be taken through how to design one.

Step 1

You will need to create a range of questions that you'd like to ask residents about their experiences of living in the local area. Using the table below, select five questions you think would be most useful in your enquiry:

Where did you live before you came here?	Would you say this area is very diverse?	What was the main thing that brought you here?
What do you consider your ethnicity to be?	What is your current household income?	How has the high street changed since you've been living here?
How has the population changed since you moved here?	How long have you lived here?	How long do you expect to live here before you move?

Step 2

Using the five questions from the table, you'll now need to decide how a resident might answer each question. For example:

Q1 How long have you lived here?			
Less than a year	Between 1 and 5 years	Between 6 and 10 years	More than 10 years

Collecting data

For each of your chosen questions, select some possible responses and fill in the blanks:

Q2 Where did you live before you came here?			
Nearby	Another part of [local area]	Another part of the UK	Another country (please tell us where)

Q3 How has the population changed since you moved here?			
There are more people from different...	There are ... young people	There are ... older people	It hasn't...

Q4 What was the main thing that brought you here?			
Family/friends			

Q5			

Step 3

The next step is to add an open question to your questionnaire. You are free to choose what kind of question you'd like to pose, so long as it is related to population change. For example:

Q6 How do you think this area will change in the next 10 years?

Step 4

At the end of some questionnaires you may find what's called a demographic monitoring section. This is important to help keep a record and check that the questionnaire has been answered by a diverse and representative sample of people. This section must be filled out by the respondent themselves, as this will help them feel more comfortable. Usually, demographic monitoring asks about people's ethnicity and gender, but can also include age, religion and sexual orientation. Some people may not wish to give an answer to this part of your questionnaire, which is absolutely fine! Although monitoring really helps, it is more important to be respectful of their choice.

Step 5

Create a box where respondents can enter their ethnicity if they are happy to.

For respondents to fill out themselves (optional)

Use this space to tell us how you would describe your ethnicity	

Activity

2 **a)** Give one reason why the questionnaire should not include answers from visitors to the local area.

b) How might a questionnaire with an older resident differ from one with a younger resident? Give one reason why.

c) Choose a sampling technique for your questionnaire: stratified, systematic or random. Look back to page 73 for a reminder of these types of sampling.

Step 6

Create a box that asks respondents to choose which of the options best describes their gender:

For respondents to fill out themselves (optional)

Male	Female	Non-binary	Other

It is important that all fieldwork is fair, representative and does not exclude anyone. You may have seen other forms that ask you to state whether you are male or female. When the ONS collects data on population it is often split into male and female to help them present it simply. However, simply splitting this section of your questionnaire into just 'male' and 'female' excludes many people in the population. When conducting fieldwork enquiries, it is important that we provide options for people who have gender identities that aren't binary – or strictly male and female. In the example above, there is a third option – non-binary – for anyone whose gender identity falls outside of strictly 'male' and 'female'. There is also a fourth option, 'other' for people who prefer neither male, female, nor non-binary.

You will use population pyramids to present your secondary data from the ONS statistics about the regional population of men and women in the UK, and migration mapping to present the results of your residents' questionnaire.

Population pyramids

Geographers use population pyramids to show the binary (male and female) structure of a population, broken up into age groups and biological sex. Since the ONS only provides binary population data, you can create a population pyramid to present it. Use the data from the ONS, provided in Figure B on page 82, to create your population pyramid.

Designing a population pyramid

Using the ONS population data on page 82, select the region that your local area falls into.

Use graph paper to create an accurate population pyramid. You will notice that the data is always to one decimal point.

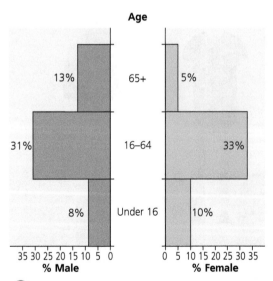

D An example of a completed population pyramid

Step 1

The first component of a population pyramid is to create the y-axis. This is where you will plot the nine age categories. Unlike other graphs, the y-axis runs through the middle of your page and has a gap in the middle that allows you to put the ages in. It will look like this:

Biological sex	West Midlands	
Male	0–9 years:	12.9%
	10–19 years:	12.3%
	20–29 years:	14.1%
	30–39 years:	12.8%
	40–49 years:	12.5%
	50–59 years:	13.1%
	60–69 years:	10.3%
	70–79 years:	8.0%
	80+ years:	4.1%
Female	0–9 years:	12.1%
	10–19 years:	11.4%
	20–29 years:	13.1%
	30–39 years:	12.7%
	40–49 years:	12.4%
	50–59 years:	13.2%
	60–69 years:	10.5%
	70–79 years:	8.7%
	80+ years:	5.9%
Total	**0–9 years:**	**25%**
	10–19 years:	**23.7%**
	20–29 years:	**27.2%**
	30–39 years:	**25.5%**
	40–49 years:	**24.9%**
	50–59 years:	**26.3%**
	60–69 years:	**20.8%**
	70–79 years:	**16.7%**
	80+ years:	**10%**

C Population survey data from the ONS

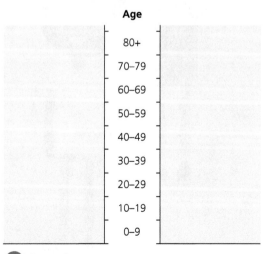

E Step 1

Step 2

The next step is to create the x-axis. This will run along the bottom of the y-axis. To the left and right, you will have the population percentage. The highest number should be close to the highest total number recorded of all the age groups. For example, in the West Midlands data, the highest total percentage was 14.1%, so the x-axis goes up to 15%. You will need to do this to the left of the middle of the pyramid and then again on the right. Label the left-hand side 'Male %' and the right-hand side 'Female %'.

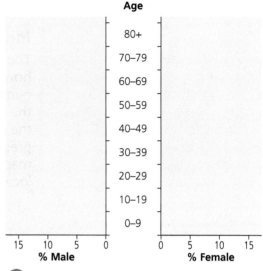

F Step 2

Step 3

The reason for needing to do this on each side of the middle of the pyramid is because it allows you to see differences between male and female data. Start by looking at the male data for the first age group, 0–9. Create a bar that extends from the middle of the axis to the left-hand side, like this:

Step 4

Now do the same but for the female data. This bar will go from the x-axis to the right-hand side.

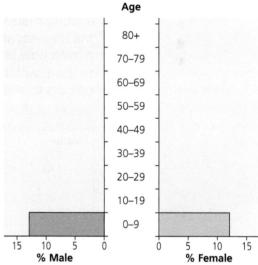

G Steps 3 and 4

Step 5

Complete the same process for each age category, so that you have nine bars in your pyramid for each gender. Add the percentage data next to each bar.

Fill in the male bars with one colour, and the female bars with another. Population pyramids are traditionally coloured in blue for male and pink for female, but be creative and choose any colours you like!

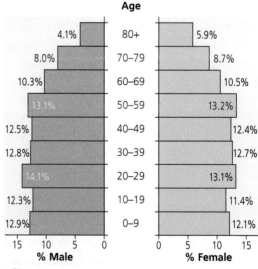

H Step 5

Presenting your data

Migration mapping

The residents' questionnaire used different questions to investigate how the population is changing in your local area. There are a number of ways you could present this information, drawing on the range of skills you have already encountered in this book. In the next activity, you will be given a number of options for data presentation, including a new one – migration mapping. Migration maps help to visualise where people who have moved into your local area have come from.

Activity

3 a) Using the data you collected or the data in the table below and an outline of a world map, show how far people have moved to live in your local area. On the map, locate the countries people moved from and use a coloured pen to draw a flow line to the UK from each of those countries. Include the direction of travel with an arrow tip at the end of your flow line. You could use different colours to show the movement of different people.

Q2 Where did you live before you came here?			
Nearby	Another part of [local area]	Another part of the UK	Another country (please tell us where)
10	10	5	Bangladesh x 2
			Greece x 1
			France x 1
			India x 1

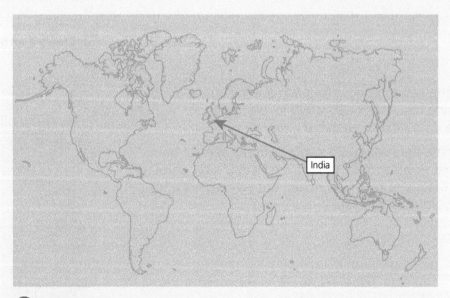

❶ A flow line showing movement of people from India to the UK

Word clouds

In your residents' questionnaire, you used a mixture of open and closed questions. While it's easy to count the number of responses to a closed question, open-ended questions are harder to present as there are often lots of different answers. One way of presenting open-ended questions is by using a word cloud (Figure J). You can make word clouds by hand using lots of colours and writing the main responses people gave to your question – for long answers, you might want to try and find a theme. For example, imagine someone gives the following response:

Q6 How do you think this area will change in the next 10 years?
It will be too expensive for me to live in.

For this response, you might choose the word 'expensive' for your word cloud. Another person may have said something different, such as:

Q6 How do you think this area will change in the next 10 years?
I think it will be far too busy, there will be too many people.

A word you might choose to use in your word cloud could be 'overcrowded'.

b) Question 1 from your residents' questionnaire asks how long a respondent has lived in the area.

Q1 How long have you lived here?			
Less than a year	Between 1 and 5 years	Between 6 and 10 years	More than 10 years
12	10	6	2

Which of the following is the most appropriate data presentation method for this question?

i) a pie chart

ii) a bar chart

iii) a radar graph.

c) Use the method you selected to present the data you collected (or the data in the table for Question 3b).

d) Look at the responses to Question 6 on your questionnaire: "How do you think this area will change in the next 10 years?" Create a word cloud like the one in Figure J to show the different answers you received for this question.

J A word cloud

Turning to your pyramids and migration map, the next step is to analyse your data.

Analysing your population pyramid

The way a population pyramid is structured reveals a lot about the make-up of your regional area. You can annotate the pyramid to show your analysis.

Activity

4 a) Look at people in the oldest age groups on your population pyramid. A high percentage of people in these groups – or a very tall population pyramid – suggests that the population is ageing, which means that the average age of a population is rising. Describe whether you think your population pyramid shows an ageing population or not, using your data to back up your answer.

b) Look at people in the youngest category (0–9 years). A wide base on your pyramid suggests a high birth rate in the population. Using your data, describe whether or not your area has a high birth rate, using your data to back up your answer.

c) People in the 16–65 years category are known as 'economically active'. Look at your population. A pyramid with a wide middle suggests that there is a high level of economically active people. Describe whether your population has a high level of economically active people or not, using your data to back up your answer.

d) Compare the male and female data. Identify one big difference in an age category between those two groups.

e) Using the table below, what might the high number in the 20–69 age category tell you? Select the correct answers from the table below and explain why.

There are a lot of young people moving into my local area.	Lots of people are in work.	There is a high birth rate.
A large number of workers from overseas may have immigrated.	A large number of workers might have emigrated from the UK.	There is a high life expectancy.

f) Using your population pyramid, annotate your answers for 4a) and 4e) onto the correct parts of the population pyramid. Figure K gives you some ideas about where annotations could go.

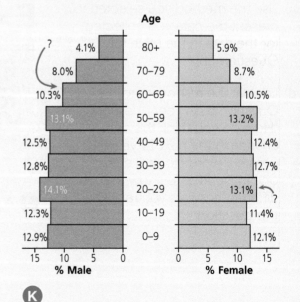

K

Comparing population pyramids with other areas

To further analyse your data, you can compare the population pyramid of your local area with a contrasting area nearby. The website Parallel gives local population pyramids of places in the UK: https://parallel.co.uk/3D/population-density. Use your data and the information provided here to help you answer questions g) and h) in the Activity box below.

Activity

4 g) Using Parallel, explore neighbouring areas to your fieldwork enquiry. Using specific examples and data from the population pyramids, compare how the population looks different from/similar to your local area in the following ways:
   i)   people under 16 (young dependents)
   ii) people aged 16–64 (economically active)
   iii) people aged 65 and over (elderly dependents)
   iv) the percentage of males compared to females.
   h) Using data, write one paragraph to describe the differences in population between neighbouring areas.

Analysing questionnaire data

The residents' questionnaire gave you some freedom to design your own questions. In this section, you will analyse some of the responses from your questionnaire.

Activity

5 a) Look at the responses to Q1 in your questionnaire, and compare these responses to the Datashine activity you did at the start of this chapter (pages 80–81). Describe whether your questionnaire reflects the Datashine map, or whether it shows something different.
   b) Explain one reason why your questionnaire data might have anomalies compared to the Datashine map.
   c) What surprised you the most about your responses to Q4, 'What was the main reason for coming to the UK?' If you did not get a chance to conduct the survey, look at the example data below to answer this question.

Q4. What was the main reason that brought you here?			
Family/friends	Work	Starting a new life	To become a UK citizen
II	IIII II	III	IIII

Activity

6 **a)** On page 82 of this fieldwork, you were provided with secondary population data for nine different regions. Identify one limitation of using regional data rather than local data.

b) The secondary data provided on page 82 was a projection from 2018. The more accurate figures would be dated 2011, at the time of the last Census. Identify one limitation of using old census data, and one limitation of using 'projected' data.

c) The secondary population data was given to one decimal place. What would happen to the total percentage if you rounded the data up or down to the nearest whole number? You may use an example to show your answer.

d) The population pyramid you created was given in age categories from 0–9, 10–19 and so on. What was one big issue with these age categories when trying to annotate your pyramid to show economically active people?

Evaluating your data

In this enquiry, you have looked at secondary population data using the ONS and the Census, created a residents' questionnaire, mapped the countries people have moved from to the UK, created a population pyramid and compared your local area to other areas in the UK. When evaluating fieldwork, it is important to reflect on the techniques used and find any limitations of them. Conducting fieldwork is never perfect, and geographers need to be able to find the limitations of their work in order to try and make it better next time.

Reflecting on population fieldwork

Population pyramids, and population data more generally, are often described as either 'male' or 'female'. However, as discussed in your questionnaire's demographic section, this isn't the most fair way of recording data on gender. To explain what we mean, have a look at the table below. It shows some examples of what hair colour people could choose when doing a questionnaire on hair products:

Q1: What colour hair do you have?				
Black	Brown	Blonde	Grey	No hair

Imagine that you have red hair, but you aren't given an option to select red hair. How might you feel? You might not be able to complete the questionnaire as your hair colour isn't included. This is not fair on people with red hair, and it is not a good example of fieldwork that is inclusive – which means including everyone and not leaving people out. Being inclusive in the way we ask questions is really important to geography fieldwork, particularly in the study of populations. There may be other ways to present your data that account for everyone in the population.

Activity

6 **e)** Think back to your population pyramid. Identify one way that splitting the population into 'male' and 'female' is not inclusive.

f) What other data presentation technique could you use to show non-binary and other gender identities?

g) Explain in your own words why ensuring your questionnaire has been answered by a representative sample of people is important in fieldwork.

h) Explain why it is good practice to ask respondents to fill out their demographic information themselves.

Creating a justification and limitations table

To evaluate your fieldwork, the next activity will give you an opportunity to justify (state why something was useful in your enquiry) and to state examples of possible limitations you experienced. You may have already thought of a few limitations from the answers provided in activity 6e)–h), so you can put them in the table too.

Activity

7 a) Complete the following table. Give at least one reason why each fieldwork technique is useful in your enquiry and one limitation of each technique.

Fieldwork technique and design	Justification: why is it useful in my enquiry?	Limitations: what problems did I experience?
Migration map	It helps visualise where people have migrated to the UK from.	We only asked for countries – it would have been better if we asked for cities or regions so that we could make our maps more accurate.
Residents' questionnaire		
Population pyramid		
Using Census data		

b) List three things you could do to improve your enquiry.

What are the attitudes towards climate change in my local area?

Learning objectives

▶ To create an environmental lifestyle survey.

▶ To design an interview.

▶ To create an infographic on local attitudes to climate change.

▶ To map the carbon footprint of one meal.

▶ To analyse interview responses.

▶ To evaluate the challenges of gathering data on food waste.

What is climate change?

The Earth has warmed up by an average of around one degree Celsius over the last hundred years. For such a small increase, you may think that it isn't anything to worry about; unfortunately, it is having many negative effects on the environment. This long-term change in the Earth's climate is known as climate change and is a real concern for a lot of people, not just geographers. Warmer temperatures mean that oceans are expanding and glaciers are melting, causing sea levels to rise. Climate change can also be linked to unpredictable weather patterns, including more frequent storms that cause flooding. These are just a few of the consequences of a warming world.

Unfortunately, humans are one of the reasons why climate change is happening. From the way we travel to the things we eat, our increasing population and lifestyle choices are making the effects of climate change worse. This is down to global warming. Around the Earth is a layer of greenhouse gases which acts like a blanket or the roof of a greenhouse. It helps to keep the Earth warm – without it, the Earth would be too cold for many living things.

Natural greenhouse gases include:

● carbon dioxide (CO_2)

● methane

● nitrous oxide.

There are several human activities that are contributing to climate change. For example:

● Agriculture – livestock farming, such as raising cattle, contributes 18 per cent of human-produced greenhouse gases worldwide. Cows naturally produce methane.

● Burning fossil fuels – many forms of transportation rely on fossil fuels. Burning fossil fuels releases high levels of carbon dioxide into the atmosphere.

● Waste – landfill sites produce carbon dioxide and methane. The bigger our population gets, the more we use and throw away. Burning solid waste produces the greenhouse gas nitrous oxide, and burning it in landfill sites leads to the production of carbon dioxide and more methane.

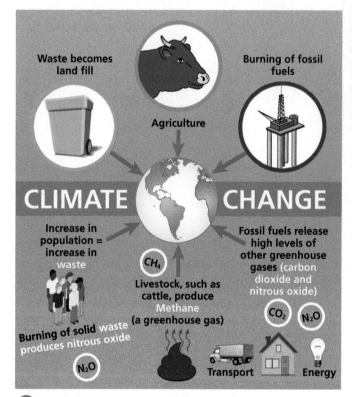

A Factors affecting climate change

What does Greta Thunberg do to help the environment?

> This ongoing irresponsible behaviour will no doubt be remembered in history as one of the greatest failures of mankind.

B Greta Thunberg

Greta Thunberg is a young climate change campaigner. She is passionate about the environment and educating people on what we need to do to fight climate change. But what sort of things does she do herself to be kinder to the environment?

Activity

1 **a)** Explain one reason why flying in an aeroplane is bad for the environment.
   b) Explain one environmental reason why Greta Thunberg chooses to eat a vegan diet instead of meat.
   c) State one example of an item that decomposes easily.
   d) Give an environmentally friendly option to complete the following sentence starters:
   - Instead of buying plastic bottles of water, I could ...
   - Instead of buying products that contain palm oil from rainforests, I could ...
   - Instead of throwing food away, I could ...

Instead of flying to the USA after a meeting in England, ...	Greta took an environmentally friendly yacht, powered by solar panels.
Instead of eating meat, such as beef, ...	Greta is vegan, which is a plant-based diet.
Instead of buying food that is harvested overseas and delivered over long distances, ...	Greta tries to eat as much local produce as possible.

Collecting data

People's choices and lifestyles are very important for finding out how much they might contribute towards climate change. In this enquiry you will use three techniques to explore some of the attitudes people have towards climate change in your local area.

Fieldwork technique 1: food waste assessment

A food waste assessment is a quantitative technique to see how much food an individual wastes across their breakfast, lunch and dinner for a period of time. It can be measured in a similar way to cloud cover – using oktas (eighths) to give a measurement of how much food is wasted on a plate (see page 41 for more on oktas).

Step 1

Give your assessment sheet a title and add the time period over which it will take place. For example, Monday 21 September 2020 – Sunday 27 September 2020.

Food waste assessment
Date:

Step 2

Underneath, write Day 1, Day 2, Day 3, etc., running down the left-hand side of the page:

Food waste assessment
Date:
Day 1
Day 2

Step 3

Split the table into four columns. In the first column, write 'Breakfast', in the second write 'Lunch' and in the third write 'Dinner'. In the fourth, write 'Total'. You can count up each eighth to calculate how much food you waste.

Food waste assessment				
Date:				
	Breakfast	Lunch	Dinner	Total
Day 1				
Day 2				

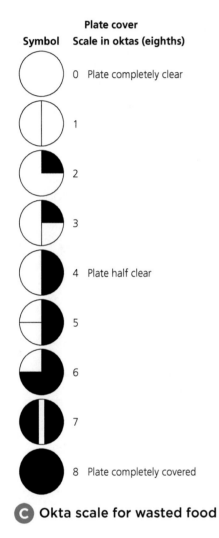

Symbol	Plate cover Scale in oktas (eighths)
	0 Plate completely clear
	1
	2
	3
	4 Plate half clear
	5
	6
	7
	8 Plate completely covered

C Okta scale for wasted food

Step 4

Add a circle into each column. To input your food waste data into your assessment, after each meal have a look at your plate or bowl and record how much of the plate is wasted food.

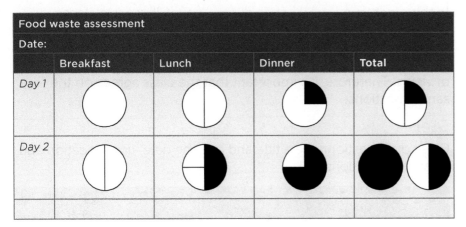

Food waste assessment				
Date:				
	Breakfast	Lunch	Dinner	Total
Day 1	○	◐	◔	◔
Day 2	◐	◧	◕	● ◐

Step 5

Calculate the total of food wasted per day. For example:

$$\frac{1}{8} \text{ plate breakfast} + \frac{0}{8} \text{ plate lunch} + \frac{3}{8} \text{ plate dinner} = \frac{4}{8} = \frac{1}{2} \text{ plate of food waste.}$$

Every year, UK households waste 4.5m tonnes of food. This amounts to around £700 for an average family with children. The UN has set a target of halving global food waste by 2030. Food waste assessments help us to understand how much food we might be throwing out of our homes every day.

Activity

2 a) Using the table below, think of five ways to reduce food waste in your household. The first one has been done for you.

Number	Idea
1	Use leftover food from dinner to make something for lunch (leftover chicken could be used to make a salad or a sandwich!)
2	
3	
4	
5	

b) Using the examples provided in earlier chapters, or your own imagination, think of two ways you might record the food waste in your household. Explain why you have chosen these two fieldwork techniques.

> We are all thinking about what we can do for the environment and this is one of the most simple and powerful ways we can play our part. By wasting less food, we are helping to tackle the biggest challenges this century — feeding the world while protecting our planet.

Marcus Glover

D A bin full of food waste

Fieldwork technique 2: environmental lifestyle survey

In this environmental lifestyle survey, you will use five closed questions with a choice of different answers in order to gather opinions. You will need to gather a total of 100 responses as a class. Using a larger sample allows you to get a wide range of views. Therefore, it's important that the class agrees on the same questions.

Step 1

Give your questionnaire a title and add the date, time, location and weather conditions.

Environmental lifestyle survey			
Date:	Time:	Location:	Weather conditions:

Step 2

Work in pairs to design five questions that relate to the following themes. Share your findings with the class:

Travel	Vegan diets	Plastic
Local produce		Recycling

E Making sure we put waste and recyclables in the correct bin is very important

Step 3

Using your five questions, create a choice of different answers that respondents can choose from. Have a look at the following example, or create your own as a class.

Q1 How often do you eat beef?			
More than 4 times a week	2–3 times a week	Once a week	Never

Step 4

Identify a popular area that you could visit as a class to conduct your survey. As a class, you will need a total of 100 responses. An easy way of doing this might be to split yourselves into 10 groups and gather 10 completed surveys per group. From this 100, you will be able to calculate percentages of your most popular responses, which will be important for your data presentation later in the enquiry. If 100 seems too many, aim for 50 – you can still work out a percentage with 50!

Fieldwork technique 3: interviews

One technique which gathers very in-depth information is an interview. Interviews generally take longer because they ask open questions. As such, you usually wouldn't do the same number of interviews as you would questionnaires. In this investigation, you will complete one interview.

One option is to interview someone who works in a restaurant. You could ask them questions about what vegan options they sell, if they get food locally, how much they waste each day, or what they are doing to help tackle climate change. A second option is to interview whoever is in charge of the cooking at home.

Step 1

In small groups, design three open questions. Keeping the questions 'open' allows the interviewee to explain their answers. An open question might look like this:

Q1 *Do you have any environmentally friendly options on your menu, such as vegan meals? If so, can you tell us why?*

You will need to write down their reply – word for word! You may find it easier to record their answer on a voice recorder, such as a mobile phone – but if not, tell them you'll be writing down their answers carefully. This usually helps them to keep their answers short.

Interview on climate change		
Location:	Date:	Time:
Q1 *Do you have any environmentally friendly options on your menu, such as vegan meals? If so, can you tell us why?*		

Step 2

Once you have designed your three interview questions, your next step is to prepare to ask your interviewee a little more about where they get their produce from. Using this information, you can create a food miles map showing how far the main ingredients have travelled to reach someone's plate.

Select an item from the menu in the restaurant or choose a meal your interviewee often prepares. Ask the interviewee the following questions:

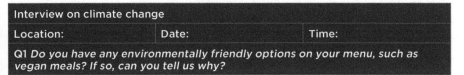

Q1 Which countries did the ingredients for this meal come from?	Q2 Do you think about where the ingredients you order come from?	Q3 Which ingredients do you tend to buy locally?

Record the meal and their answers on your interview question sheet.

Activity

3 a) Explain one reason why wasting food is bad for the environment.

b) Suggest one practical way of reducing your food waste.

c) Describe two advantages of asking open questions in interviews rather than closed ones.

Presenting your data

Using the three fieldwork techniques on pages 96–99, the next step in your enquiry is to present the data you have collected.

Creating an infographic

In this enquiry, you will map, calculate and visualise your findings into an infographic poster. An infographic is a way of visually capturing people's attention, using key bits of information to show some of the key insights from your enquiry. In this section, you are going to create some insights for the infographic.

Activity

4 Over the course of seven days, you mapped the amount of food wasted and created a total.

My food waste this week:						
Day 1	Day 2	Day 3	Day 4	Day 5	Day 6	Day 7
◯	◯	◯	◯	◯	◯	◯

a) Draw your total amount of food waste over a week in oktas. For example:

My food waste this week:						
Day 1	Day 2	Day 3	Day 4	Day 5	Day 6	Day 7
◓	◓	◯	◯	◯	◯	◯

b) Calculate the total amount of food wasted by your class this week. For this task, your classmates will need to share the total amount of food they each wasted.

c) Using the total amount of food your class has wasted, calculate how much this would be over the course of a year. For example:

- Whole class = 3 and ²/₈ plates of food in a week.
- ¹/₈ = 0.125 so ²/₈ = 0.25
- 3 + 0.25 = 3.25
- Number of weeks in a year = 52
- 3.25 x 52 = 169 plates of food a year!

Collating survey responses as a class

Collating all the survey responses as a class is a great way of getting lots of data for your enquiry. The more data you collect - sometimes called the **sample size** - the more confident you can be of your results. In the next activity, you will gather together all the data you collected as a class.

Activity

4 d) Create tables to show the total survey responses collected as a class. Use the following examples to help you:

Jo's group:

Q1 How often do you eat beef?			
More than 4 times a week	2–3 times a week	Once a week	Never
5 people	2 people	3 people	0 people

Class responses:

	Q1 How often do you eat beef?			
	More than 4 times a week	2–3 times a week	Once a week	Never
Jo's group	5	2	3	0
Group 2	4	3	2	1
Group 3	4	3	2	1
Group 4	2	0	6	2
Group 5	2	3	5	0
Group 6	1	5	4	0
Group 7	1	0	7	1
Group 8	2	4	3	1
Group 9	1	5	3	2
Group 10	2	6	2	0
Total/100	**24**	**31**	**37**	**8**

	Q2 Have you ever chosen to take a train or boat rather than fly for a holiday overseas?		
	Yes, often	Once or twice	Never
Total/100	8	30	62

	Q3 How often do you buy local produce?		
	Regularly	Occasionally	Never
Total/100	35	51	14

	Q4: Do you use a refillable water bottle?		
	Always	Sometimes	Never
Total/100	18	71	11

	Q5 How often do you use your recycling bin at home?		
	Regularly	Occasionally	Never/We don't have one
Total/100	51	39	10

Presenting your data

Mapping food and visualising interview answers

In your final fieldwork technique, the interview, you asked open questions to someone in your local area. You also investigated how far the food from one meal travelled to get onto a plate in a restaurant, or at home. One way to help present this data is to make the data come to life in the infographic through food mapping and using interview quotes.

To map the food miles of a meal, you will need an outline of a map.

Activity

5 **a)** Using the data you gathered for the origin of the ingredients, map each one by using a colour pen to draw a line from the country of origin to the UK.

b) Write the name of the country next to the mark you made on the map.

c) Using Google Maps, calculate the distance that each item has travelled. (If you are using the dataset provided, complete the table below.)

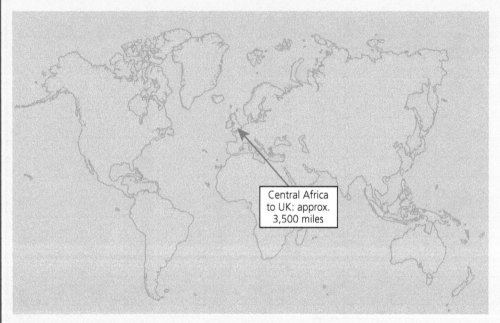

Central Africa to UK: approx. 3,500 miles

F A food mile map – for one ingredient

Spaghetti bolognese		
Ingredient	**Country of origin**	**Approximate distance from country of origin to UK**
Pasta (spaghetti)	Italy	
Tomatoes	Israel	
Garlic	England	
Red wine	France	
Basil	Central Africa	3,500 miles
Onion	England	

G Food miles for spaghetti bolognese

Activity

5 d) The final piece of information to include in your infographic is going to be a quote from your interview. Using the notes from your interview, find one or two short quotations that you found particularly interesting. Here are some examples:

> We usually throw out about a bag of food waste a day – but we always put it in a compost bin, and never into a regular bin.

> I should bring in more local asparagus, but it is only in season for about two weeks in the UK – I can get it almost all year round from places like South America.

> I don't really offer vegan options because they don't sell as well.

e) The final touch is to put all of this together onto an infographic poster. Combining the food waste data, the statistics from your survey, the food miles map and the interview quotes, your infographic will be bursting with important information. Title the infographic clearly with the enquiry question – and be as creative as you like. Make the numbers and data as big and colourful as you like to make them stand out.

What are people doing in my local area to tackle climate change?

In our class, around **3.25** plates worth of food was wasted in a week  That's **169** plates of food in a year!

25% of people we surveyed said they ate beef more than **4** times a week!

 This meal travelled over **8,000** miles before it reached our plate!

H A geography of food infographic

In order to tell your geographical story, you first need to analyse the data you have collected.

Analysis

Using the calculations from your survey and food waste assessment, and the data from your interview, work through the following activities.

Activity

6 a) With your total class food waste, calculate the average daily food waste per student.

b) Carefully analyse your class climate change survey data. Using the percentages as examples, compare the responses to each question. For example:

> '... of people we surveyed did own a recycling bin at home. However, only ... % of people regularly use their recycling bin.'

> 'We asked 100 people in our local area if they had ever taken a train or boat to travel overseas. Only 8% of people had taken an alternative to flying, compared to 62% of people who fly. That's almost 8 times as many people flying as travelling by boat or train to go on holidays.'

c) Look at your food miles activity. Calculate the total distance food has travelled for one meal. Use the spaghetti bolognese example if you like.

d) Which item travelled the furthest distance?

e) Using the internet, research the locations that your answer to 5d) could have also come from. Can you identify any ingredients that could be sourced in the UK, but weren't?

f) Give one example of how the person you interviewed *was* being environmentally friendly.

g) Give one example of how the person you interviewed was *not* being environmentally friendly.

I Some households use food bins so their food waste can be composted

Evaluating your enquiry

Evaluating how your enquiry is designed and how data is collected is a key part of your fieldwork. This might mean thinking more carefully about the way you pose questions, or the units of measurement you used. The following activities will question how you conducted your fieldwork enquiry.

Activity

7 **a)** What was challenging about recording your food waste scores by using eighths?

b) How could you make the food waste assessment more accurate? Describe one alternative suggestion.

c) When choosing 'closed' answers to the question 'How often do you eat beef?', give one reason why option 3 might be better than options 1 and 2.

Option 1:

Q1 How often do you eat beef?		
All the time	Occasionally	Never

Option 2:

Q1 How often do you eat beef?		
More than 4 times a week	2–3 times a week	Once a week or less

Option 3:

Q1 How often do you eat beef?			
More than 4 times a week	2–3 times a week	Once a week	Never

d) Give one challenge you experienced doing the fieldwork for the food miles activity.

e) Give one advantage and one disadvantage of completing the interview rather than a questionnaire.

f) Throughout this fieldwork guide, you have explored nine different geographical enquiries. In this final activity, link the previous eight enquiry themes to climate change. How do they impact on climate change, and how might climate change impact on them?

Cycling	Sustainability	Businesses
Microclimate	Coasts	Flooding
Quality of life	Population	Landscape

Congratulations! You have completed all nine geographical enquiries in this book. I hope you have enjoyed learning all about fieldwork and are ready for your next adventure.

Glossary

Ageing A term used to describe a population where the average age is rising

Anemometer A fieldwork instrument used to measure wind speed

Annotated map A way of presenting data on a map by adding labelling information onto it

Anomalies Any unusual results in the data

Aspect The direction a building or place is facing

Attitudes How people feel about a project or proposal

Axis A line that runs through a rose chart or on a graph to show different values

Backwash Seawater moving back into the sea after a wave

Bar chart A graph that shows frequency data as blocks or bars along the axes

Business Any organisation that makes goods or provides services

Business decision-making exercise An exercise used to determine what kind of business to open in a given area

Business location survey A quantitative survey used to score the best location in which to open a new business

Census A complete population count for any given place or area. Census data has been collected in the UK since the 1800s. The Census is updated every ten years.

Choropleth map A map that uses different shades of a colour to show different values or numbers

Climate change A long-term change in the Earth's climate, including fluctuations in temperature

Closed questions Questions that require a 'yes' or 'no' answer, or a tick in a box

Clustered bar chart A bar chart that displays more than one set of data next to each other

Coastal management The different options available for defending a coast. Often, they are based on building structures to protect cliffs and beaches.

Coastline the strip of land that forms the boundary between the land and the sea

Coding A technique used to put data (often text or writing) into different categories, to help look for common themes

Commercial properties Buildings that are used for business

Compound bar chart A chart that stacks bits of information on top of each other

Data The information you gather when doing fieldwork

Demographic monitoring A way of capturing information on diversity in questionnaires

Deposition Laying down material to form new land

Development The process of change by which people reach an acceptable quality of life

Economy The wealth and resources of a country in terms of the goods that are produced and consumed there

Egan wheel A tool for investigating sustainability

Employment sector All the types of jobs in the economy grouped into sectors

Environmental quality survey A bipolar quantitative survey used to examine the quality of the environment, such as traffic and buildings. Bi-polar means that it ranges from a minus number to a plus.

Erosion The process by which rocks and soils and materials are worn down and moved elsewhere due to mechanical and chemical action (such as wave power, or salts in water) or weathering processes (such as wind, rain, plant roots, etc)

Ethnic group A category of people that share the same language or culture

Fieldwork   How we study geography outside of the classroom. Fieldwork is completed using some key steps: preparation, collection, presentation, analysis and evaluation.

Fieldwork technique   The different ways you gather your data

Flood   An area of land saturated by water due to overflowing rivers

Flood hydrograph   A graph that shows how a river responds (the height or volume of water) after a rainfall event. Sometimes the rainfall is shown on the same graph as well.

Flood map   A map produced by the Environment Agency to show areas at high, medium or low risk of flooding

Flood plains   An area onto which a river floods, usually in the lower course

Flow rate   The volume of water or fluid flowing through an area

Food miles map   A type of data presentation that visualises where food has travelled from to reach our plates

Gender identity   How people feel about their gender. This could be male or female, but it could also be neither of these

Global warming   The increase in the Earth's temperature

Groyne   A barrier built into the sea from a beach to help prevent erosion

Hard engineering   Coastal defences constructed from concrete and steel

Hazard   Something in the environment that could cause harm

Infographic   A visual representation of data, including pictures, quotes and statistics, that can be made into a poster

Interception   When trees or plants collect rainfall before it reaches the ground

Interview   A qualitative fieldwork technique that gathers in-depth information on people's opinions

Key enquiry question   The question guiding your fieldwork enquiry

Landscape management   How people use the land, i.e. whether it is farmed, woodland or built upon

Limitations   Problems you might have found during your fieldwork which could have affected the results

Line graph   A graph in which data is plotted as points along the axes and then joined together with a line. These graphs are used to represent a particular trend over time

Local sustainability survey   A fieldwork technique that maps the sustainable characteristics in an area that help the planet

Longshore drift   The zigzag motion of sediment being transported along the coastline as a result of winds blowing the waves at an angle to the beach

Microclimate   The local climate of a small-scale area

Migration mapping   A type of data presentation that shows where people have migrated from into the UK on a world map. This is the opposite of an emigration map, which shows where people have left the UK to settle

Non-binary   When something does not involve only two things, or two options. This is a term that is often used to describe gender that does not fit into male or female (binary) categories

Okta   A unit of measurement to show cloud cover. It is measured in eighths of the sky and is presented as symbols.

Open questions   Questions that allow respondents to explain in more detail

Pictogram   A creative form of data presentation that uses symbols or icons to show data

Polygon   The irregular shape created by the data plotted in a rose chart

Population   The number of people living in an area

Population pyramid   A graph that shows the distribution of a given population by age and biological sex

Primary data   Information collected by someone first-hand, using fieldwork techniques

Qualitative data   Data that uses opinions or descriptions rather than numbers

Quality of life The general wellbeing of individuals and society. It is measured by looking at areas such as housing, transport, green space and safety.

Quantitative data Factual data that can be counted

Questionnaire A set of questions to ask people about their lives or opinions

Radar graph A chart that shows quantitative data along different points to create a shape. Also called a rose diagram

Representative sample A sample that closely represents or reflects the wider population

Residential properties Buildings that are peoples' homes

Residents' questionnaire A type of questionnaire aimed at people who live in a particular area, to find out their views on an issue like population change

Respondents The people who answer your questionnaire or survey

River discharge The volume of water per second that is flowing down a river. Usually measured in cumecs or cubic metres per second.

Sample size How many people or how large an area you studied in your fieldwork. For example: the number of questionnaire responses (sample size = number of respondents) or the size of the area by the river you studied.

Scale A scale on a data collection sheet is a sliding list of numbers, like 1 to 5 or +2 to -2, that geographers use to help them score things

Scatter graph A type of graph that shows relationships between different sets of data

Sea wall An artificial barrier built to stop the sea eroding the land

Seafront properties Buildings on the coast, facing the sea

Secondary data Data that is from someone else's study, for example a flood map of an area. This could be from many years ago, or it could have been produced recently.

Sources of error Mistakes that might happen during data collection

Subjective Based on personal opinion. This often means results can vary significantly.

Sustainability Using resources and materials in a way that will balance the needs of the present without making things worse for the future

Sustainable community A place where people want to live and work, now and in the future

Sustainable community assessment A sustainability survey based on the Egan wheel

Sustainable management This is sometimes called soft engineering and is coastal management that is designed to last a long time. It normally has a lower impact on the environment.

Swash Seawater moving up the beach after a wave

Traffic count Used to count the number and type of vehicles going past within a set time

Transport Movement of material (along the coast by waves)

Trend A general direction in which something is developing or changing, i.e. going up or down

Urban heat island effect The dome of warm air that builds up over towns and cities

Wind tunnel effect When wind picks up speed between two physical features close together, like buildings

Word cloud A way of showing the different answers to open-ended questions. Words that are bigger in a word cloud mean that lots of people gave that particular response